The whole truth about ECONOMY DRIVING

by Doug Roe

Editor:	Carl Shipman	Library of Congress Cataloging in Publication Data	ISBN: 0-912656-22-0
Publisher:	Bill Fisher		Library of Congress Catalog Card No. 74-15212
Photography:	Doug Roe	Roe, Doug.	H. P. Books No. 22
	John Thawley	The whole truth about economy driving.	
Book Design and Assembly:	Josh Young	1. Automobiles—Fuel consumption. 2. Automobile driving. 3. Automobiles—Maintenance and repair.	H. P. Books, P. O. Box 5367 Tucson, AZ 85703 602/888-2150
	Nancy Fisher		
Typesetting:	Ellen L. Duerr	I. Title. II. Title: Economy driving.	
Front Cover:	Josh Young	TS154.R63 629.22'22 74-15212	© 1975 H. P. Books

CONTENTS

Page 4—Chapter 1
A PENNY SAVED...

If you want to save bucks, you can.

Page 6
HOW TO CHECK
YOUR GAS MILEAGE

Begin by finding out how bad it is.

Page 10—Chapter 2
ECONOMY DRIVING TECHNIQUE

Anybody can learn to drive for economy.

Page 15
BASIC RULES FOR
ECONOMY DRIVING

Five simple ideas you can put to work.

Page 29
WOULD YOU BELIEVE
376.59 MILES PER GALLON?

Shell Oil Company experts prove it's possible—but you won't go that far.

Page 39—Chapter 3
ECONOMY ON YOUR
VACATION TRIP

Maybe when you get there, you'll still have enough money to drive back home.

Page 55—Chapter 4
SAVING IS NOT
SPENDING SO MUCH

Good advice for everyday savings on your car expenses.

Page 77—Chapter 5
RIP-OFFS

Saving is not spending anything!

Page 93—Chapter 6
BUYING A CAR FOR ECONOMY

How to choose the best car for you.

CONTENTS

Page 103
ACTUAL TEST DATA

	SAE City	SAE Highway	EPA City	EPA Highway
dsmobile 98	10.1	15.2	11	15
dsmobile Toronado	10.6	15.2	11	16
ymouth Fury Salon	10.2	16.8	16	21
ymouth Gran Fury	9.0	16.3	12	17
ntiac Catalina	10.8	17.4	12	15
ntiac Grand LeMans	10.1	16.6	13	18
ntiac Grand Prix	10.0	15.7	12	17
yota Corolla	20.2	27.5	21	33
MC Gremlin 8	15.6	19.4	19	24

Cold facts from the U. S. government and independent Union 76 Economy Tests.

Page 107
USED OR ABUSED

Maybe the little old lady ate the cream puff, but here's a check list for buying a used car.

Page 113
HOW MUCH DOES A CAR COST?

CAR	YEAR 1 14,500 miles	YEAR 2 13,000 miles	YEA 11,50
Standard four-door sedan Price: $4,251	$2,427.27 16.7¢ per mile	$1,807.03 13.9¢ per mile	$1,82 15.9¢ mile
Compact two-door sedan Price: $2,910	$1,570.14 10.8¢ per mile	$1,432.19 11¢ per mile	$1,38 12.1¢ mile
Subcompact two-door sedan Price: $2,410	$1,283.37 8.9¢ per mile	$1,165.10 9¢ per mile	$1,09 9.5¢ p mile

Official U. S. Department of Transportation info—but you have to add more for inflation.

Page 115
SELL IT YOURSELF

A three-dollar classified ad can get top money for your old car.

Page 117—Chapter 8
A SHORT COURSE IN MECHANICS

A simple explanation of how your car works—so you can learn to work on it.

Page 124
HOW A CARBURETOR WORKS

A full anatomical description including location of the gizzard.

Page 138—Chapter 9
LEARN TO TUNE-UP YOUR CAR

Do it yourself—save money and enjoy a good-running car all the time.

Page 149
FROM CHAMPION SPARK PLUG COMPANY

Learn to "read" your spark plugs so you know what's happening inside your engine.

Page 164—Chapter 10
INSTRUMENTATION FOR ECONOMY

Sophisticated test equipment plus some things you can use in your car.

Author Doug Roe has been professionally engaged in automotive engineering, testing, tuning and economy for over 25 years. Economy-run winner, Detroit test engineer and carburetion specialist, he now heads an independent automotive engineering and consulting firm.

1 A PENNY SAVED IS NOT WORTH IT?

If you think it's no big deal to save a penny, or a nickel or dime—if you think economy gets to be a worthwhile proposition only when it lays ten bucks in your hand or leaves a hundred in your checking account—welcome. You've come to the right book, friend. We've got to *change your way of thinking.*

The gambling casinos in Las Vegas operate on a percentage. At some game such as roulette the house will win about 5% of every bet you make—on the average. The management is happy to collect 5% of the minimum you are allowed to wager, and they are happy to accept 5% of a thousand-dollar bill. Their main goal is to keep a lot of people betting a lot of times and just let the percentage put bucks in the back room.

That's a very good way for you to think about making economies in your life—car, house, food, clothing or whatever. Think of the percentages, not the pennies. Ten percent is ten percent. If you dribble away ten thousand dollars in a year and you start saving ten percent on everything, or on the average, you'll wind up the year with a thousand dollars in your jeans.

Sometimes your saving will only be a penny, sometimes it will be a dollar or more. Don't count the pennies. Count the percentages! When you start thinking that way, you have tuned your mind to practice economy. That simple idea put to work has made a lot of people rich.

THE LAND OF GOOD INTENTIONS

Nearly everybody intends to do right—starting tomorrow or the next day. We intend to get the front wheels aligned before they chew up another pair of tires. We intend to go to the dentist before little problems become a major expense. We intend to put some money in the savings account and leave it there.

Most people live in the Land of Good Intentions and never leave. If you live there, you should move.

THE LAND OF DOING SOMETHING

A few people actually do some of these things and make a great discovery! The more you do, the easier it gets. When you have made some financial progress or when you have done something that puts you more nearly in charge of your life you begin to feel virtuous. You become a stronger person and think in a more practical way. You begin to see that it's really possible for you to have more financial security, and have some money to do something that's really important to you.

GETTING STARTED

The main thing is to make a start. Reducing your cost of transportation is a

The economy runs of the 1960's were taken seriously by competitors but the general public considered them an unexciting cross-country race or else harmless entertainment. Photo courtesy of Mobil Oil Company.

very good way to begin. You spend a lot of money on cars, insurance, operating expense and repairs. It's easy to save ten percent of that money. More if you want to. This book will show you how.

GETTING ACQUAINTED

Writing this book began when the nation started to feel the energy crisis—late 1973. The experience and know-how you will find here goes clear back to the economy-run days of the late 1950's and early '60's. The current energy crisis hit the public in combination with runaway inflation. Thus it is even more important to review *all aspects* of saving fuel and money.

The test data appearing in this book was generated by actual tests to get the information first hand. There is no more accurate data available to the driving public than that appearing in this volume.

I tried to determine what's best for you with today's legal requirements, technology and everything else that concerns the 1970 and later cars. The book touches on vehicles older than 1970 because many are still running and people are trying even harder to get more service from them. As operating costs increase, more drivers will turn to older vehicles—something for you to keep in mind as a way of reducing the cost of operating a motor vehicle.

A lot of this book is devoted to getting better gas mileage and a lot is devoted to the basic goal of spending less money on a motor vehicle—better mileage is just one of the ways you can save money. There are others.

Some of these ways to economize will be practical for you to do—some may not apply to your talents or preferences. One thing everybody can do is modify driving habits to get better gas mileage. You'll be surprised at the payoff.

HOW TO CHECK YOUR GAS MILEAGE

Checking gas mileage gets to be a real bore after two or three tanks of gas, but if you are really interested in economy you will keep track of gas mileage either continuously (long-time or cumulative method) or just occasionally (tank-by-tank method) for a particular purpose. There are two ways to check mileage. The tank-by-tank method is what you should use if you plan to tune your car for better mileage or plan to try out some of the driving techniques in the next chapter.

Then came the fuel shortages and long lines at gasoline stations. George J. Public finds economy less fun—more necessary.

Union 76 conducted fuel economy tests of the 1975 car models under strictly controlled conditions at Daytona Speedway. Yes, friends, it's time to start fuel economy.

Built into the speedometer is an *odometer* which registers total miles driven. Some cars also have a resettable or "trip" odometer which can be set back to zero and is handy for checking mileage from a tank of gas.

A minor bit of organizing helps a lot when checking mileage. Put a clipboard or notebook in your car and use it. Writing the odometer reading on the back of your charge-card receipt, and then losing the receipt, is unscientific.

Here are some names you need to know. The *speedometer* tells you how fast you are going. Built into the speedometer is another instrument, called an *odometer* which registers the total miles traveled by the car. The odometer appears to be part of the speedometer and most people don't know what to call it.

Some cars have two odometers—one to accumulate the total miles the car has traveled and another which can be reset to zero. A resettable odometer is often called a *trip meter* because you can set it to zero at the beginning of a trip. A trip meter is handy for checking gas mileage but not necessary. Without one, you just write down the mileage shown by the odometer at the beginning and the end of a test period and subtract the smaller number from the larger.

Don't worry about the odometer and speedometer being in error. Don't try to correct odometer error and compute gas mileage at the same time.

Try this: Motor over to your local service station and fill the tank. YOU fill the tank. Let the attendant do the windshield or watch girls while you fill the tank to within a couple of inches from the top. Somewhere in that area of the filler neck there will be a dimple, or ledge, or heavy scratch which you can use as your *full* mark. Write down the mileage, including the tenth-of-a-mile figure at the far right on the odometer.

The next time gas is purchased, YOU fill the tank again. Make certain the car is level each time you fill up. Bring the gas level in the tank right back up to the previously-selected mark and make note of how many gallons and tenths it took to fill up. Divide the number of gallons into the number of miles driven.
Example—57.8 miles divided by 4.6 gallons gives 12.56 miles per gallon.

This tank-by-tank method allows you to make tuning changes, or test your driving habits without running through a lot of gas to see if you are making any improvement.

Let's say you really don't care about checking mileage for every tank of gas but would like to know what your gas mileage is over a long period—say a month.

Here's how: Fill the tank and note where the level of the gas is in the filler neck. Record the odometer mileage—including the tenths. After that, you need not be particular about the tank being full or you can even buy five or ten gallons and not a full tank.

The important thing is to keep a record of all gasoline that goes into the tank. Let the attendant put in the gas if you wish, but you should personally check the pump to see the number of gallons pumped.

When you end the test, you fill the tank and bring the level back up to your special mark again—where it was when the test began. Now total all of the gallons of gas added in the test period. Check the odometer and figure out how many miles you have driven during the test. Then divide the number of miles driven by the number of gallons consumed to figure the miles per gallon the car averaged during the test period.

Either way, the accuracy of the information you end up with depends on the accuracy of your records. Keep a notebook or clipboard in the car with a pen or pencil so all the information can be recorded right there in the driveway of the service station. The deeper you get into the project of getting good gas mileage, the more you will come to realize that there is no magic to it—just a lot of interesting work. Part of that work is keeping very accurate records of everything you do and what effect it has on gas mileage.

Earlier I mentioned speedometer and odometer error—don't worry about that. The 12.56 figure might not be an accurate one because of speedometer error, but if you continue to use that as a base, then the subsequent mileage figures taken will tell you if progress is being made. That's the important thing.

USING BASELINE DATA

The idea of a base, or baseline, is useful in any kind of testing. It's used at auto proving grounds, in setting up for a fuel-economy run, and in most other kinds of testing. If you are checking gas mileage, the first thing you have to find out is the exact gas mileage your car is getting in its present condition and with your normal way of driving. Measure it and find out. Write down the result and the date in your notebook. That figure becomes your baseline.

Then when you try something different, you can compare your new gas mileage with your baseline. If you find an improvement and decide to keep on doing it the new way, then you have a new baseline to work from in the search for more improvement.

Another simple idea that doesn't occur to people who are not accustomed to test operations is this: Test only one change at a time. If you try different tire pressure, a different carburetor setting, and keep your speed under thirty miles per hour (MPH) all at the same time, you won't know which thing had what result.

In the good-ole days, when economy was for fun, some of the fun was at the finish. Detroit, 1963.

2 DRIVING TECHNIQUE— ANYBODY CAN DO THIS!

The smallest available engine is not always best for economy. It depends on what you haul and how you haul it.

I was in the thick of the highly-publicized economy runs in the 1950's and 1960's when John Public really believed in cars getting 160 miles per gallon (MPG). Mr. Public thought he was being denied this gas mileage by a conspiracy between auto makers and oil companies. The public pointed to the economy run results as proof and continued to nourish those dark suspicions.

Meanwhile, the auto companies took the rather ridiculous posture of full-page advertising of the results—everybody won something—coupled with a hard sell of models which came nowhere near the advertised mileage triumphs.

In those days, I participated in the economy runs as driver, training coach for other drivers, factory engineer, perpetrator of little tricks on other competitors.

Today there is a rebirth of interest in fuel economy and an agency of the government—the Environmental Protection Agency—is somehow in the business of testing cars for gas mileage and publicly announcing the results. Once again the car makers are advertising the results—when they seem favorable—and complaining bitterly when they disagree with EPA's figures.

The economy runs are being reborn. There was a Union 76 Economy Test program at Daytona Speedway for the 1975 models. I was there as a team manager and driver of seven of the 82 vehicles entered. I took first place mileage in four out of five classes in which I ran, third place in another class, and a teammate from my firm took another first. One of my economy students, Paula Murphy, won several classes.

IS 160 MPG PRACTICAL?

No, it isn't. It's *possible,* but not practical except to get headlines. You wouldn't want the car that can do it and you wouldn't want to drive the way such a car has to be driven.

But you are probably figuring that 8 to 12 MPG isn't very practical either in terms of the price of gasoline these days, the impact on your wallet, and the continued depletion of the world's natural resources.

There is a middle ground that offers you worthwhile savings without requiring you to drive around towing a constant traffic jam behind you. You can use some of the driving methods of the economy-run champions along with your own common sense. That's what this chapter is about.

THE PAYOFF

There are some real payoffs from doing this; some put money in your pocket—some aren't measured by money but may be equally important. The obvious saving is in the cost of gasoline, but there are several others:

Driving Becomes Less Boring—That's hard to believe until you actually experience driving for economy. You probably think that tootling around nursing each drop of fuel is a very boring way to get from here to there.

But answer this question. Isn't most of your driving rather boring anyway? The daily roundtrip to work isn't a thrilling adventure is it? You don't see many commuters laughing and giggling along the way, on account of the joy of fighting traffic and the high cost of transportation.

How 'bout your vacation trip or the weekend jaunt to Aunt Minnie's so the kids can smell that country air? You probably do most of the trip on big highways which are dull—all the way. New highways are carefully designed to simplify the task of driving and reduce it to hours of sitting there, looking at the

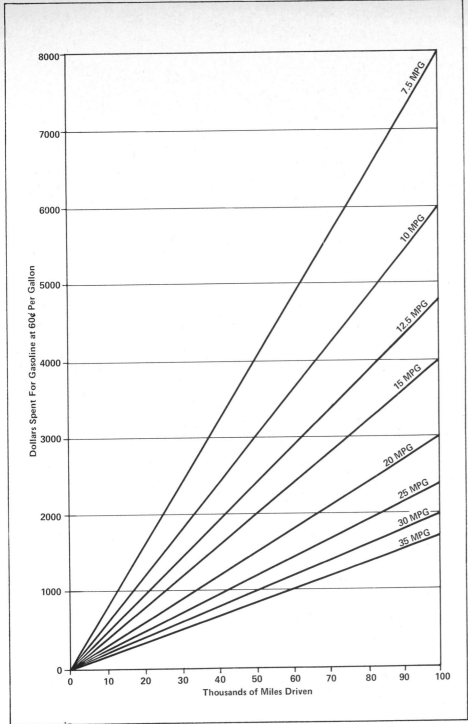

You can save bucks by driving less or by using less fuel per mile.

white line. That's boring!

What do your passengers do to pass the time? The kids enjoy the trip by fighting or beating on each other in the back seat. Or else everybody goes to sleep. Except you. You just keep on staring ahead.

For fun and profit, try converting all your routine driving into a contest between your clever scheming intellect and the forces of evil which want to deny you good gas mileage. Now you have something to do besides the routine and you get a payoff from it. When you start playing the economy game, you'll find it can be interesting.

You Save Money on Gasoline—Yes, you do. The chart on this page shows how much you can save.

You Have Fewer Car Repairs—One basic element of economy driving is smoothness. No jackrabbit starts or screeching stops. That makes the parts of your car last longer. You will pay for fewer tires, fewer brake jobs, fewer transmission and engine rebuilds. If you postpone a transmission and engine overhaul until somebody else owns the car, you have saved a thousand bucks!

You Don't Get Smashed and Killed as Often—If you are thoughtful enough to be even mildly interested in a book like this, I know something about you. You are not a wild reckless driver. You have a good driving record already and insurance companies like to have you aboard.

You manage your own driving rather well, from a safety standpoint, and are very watchful of the other guy. Most of your traffic problems result from something the other guy does, rather than an error on your part.

It's interesting that economy driving methods go hand in hand with good safe driving habits. For economy, you will watch traffic closely and plan your moves well ahead of time. You will be aware of road conditions, hills and turns, stop signs and the total environment around you. These are the same things you do for safety. Now you have two reasons to do them and you get a double payoff.

WHERE DOES THE GASOLINE GO?

Before getting into driving methods, it may help to spend a minute learning what a car does with the gasoline you put in the tank. You know the engine

Nobody claims this is exciting motoring. But it's slightly less boring if you occupy your mind by plotting to save a drop or two of gas.

In a couple of hours this straight empty road may angle to the left. Hold a steady legal speed and get some fuel economy along the way.

Speed MPH	Increased Wind Resistance	
30	(baseline)	1.00
40		1.77
50		2.77
60		4.00
70		5.44
80		7.11
90		9.00
100		11.11

Considering 30 MPH as the reference or *baseline,* wind resistance is four times as much at 60 MPH.

burns it and the result is mechanical energy to propel the vehicle.

When you pay for gasoline, you are actually buying some energy in the same way that your electricity bill pays for the electrical energy you use around the house. If you are economizing on electricity, you try to use less of it. To use less gasoline, it helps to know how the energy is used in operating an automobile.

Acceleration—We commonly use the word *acceleration* to mean increasing speed—from standstill to driving speed, or from a slow speed to a higher speed. Your car uses more gasoline to accelerate to any speed than to cruise at a constant speed.

Wind Resistance—Once you reach a desired cruising speed, you use less gas to maintain that speed. Your right foot will readily confirm that observation.

In the conditions of outer space—no friction and no wind resistance—a body at some speed will theoretically travel at the same speed forever and require no energy at all to do it.

The reason it takes energy to maintain cruising speed here on earth is due to wind resistance and several varieties of friction.

Wind resistance is the major factor at higher speeds. In fact, wind resistance goes up 4 times if you double cruising speed and it goes up 9 times if you triple cruising speed. Here's a table to show the increase in wind resistance, with 30 MPH as the baseline.

There's an old saying among racing engine mechanics that "speed costs money." It costs in more ways than just the price of the high-powered engine, as you can see.

Rolling Resistance—Even if there were no wind resistance at all, it would still require some energy to cause the car to roll along the highway. This is due to friction in the engine, in the gears and in places such as the wheel bearings.

Also, there is rolling resistance due to the type of tires you use, the amount of air pressure in the tires and the surface they roll across. If you have ever tried to push a vehicle—even a bicycle—in sand or soft dirt, you appreciate the effect of surface. Tires rolling on wet pavement have increased rolling resistance because they are pushing water. Economy suffers dramatically as a result.

An unusual type of energy requirement comes from the fact that the tires deflect a little bit and are flat on the bottom where they contact the paving. As the tire rolls along, a different part of the tire becomes flattened. Just to deform the tire takes some force and energy. If you don't believe that, try doing it with your hand.

A steel railroad wheel rolling on a steel track is about as good as you can do in reducing rolling resistance due to tire flexing. Don't invent steel wheels for automobiles though because the highway department won't let you use them. They would destroy the paving.

You can reduce rolling resistance by using higher tire pressures to reduce the amount of tire flexing. But if you carry that bright idea too far, you will wear out the center of the tire tread quickly and spend money a different way.

Uphill and Down—It takes extra gasoline to get a car up a hill because—in addition to overcoming all the forces listed so far—the engine must also *lift* the car to a higher altitude. Some energy is *stored* in the car just because it is now at the top of a hill. Which only means that it can coast back down the hill so it will give you back some of the energy used in putting it up there in the first place. You never get it all back though because rolling resistance and wind resistance are taking energy all the time, both uphill and down.

If you drive the same distance on flat terrain and on hilly terrain, you will get worse gas mileage in the hills and dales.

Engine Efficiency—The engine burns gasoline and uses the energy to conquer the obstacles discussed above. Unfortunately the engine itself is a great energy waster because it does not convert all of the energy in the gasoline into mechanical force to drive the car. Much is wasted even under the best conditions.

If you allow an engine to get out of tune or misfire, then it gets even worse.

Emission engines have purposely been built to be inefficient. The controls which

Climbing a hill takes gasoline energy to *lift* the car up the hill. You can get some of the energy back while coasting down the other side. But never all of it.

This bar graph shows how gasoline mileage improves as the engine gradually warms up from a cold start on a cold morning with the air temperature at zero degrees. If you drive 2 miles to work and then park, you got about 2.5 miles per gallon. Even at 14 miles, fuel economy is only about half as good as it would be when the engine is fully warmed up.

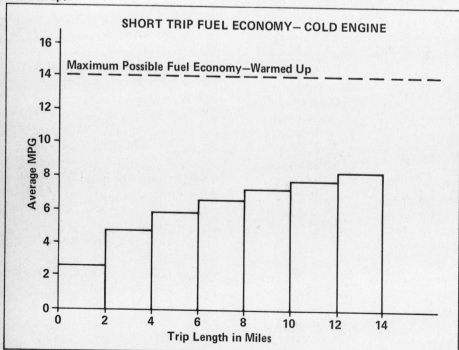

are hung on modern engines to help reduce emissions also reduce efficiency and worsen gas mileage.

FEATHERFOOTING IS NOT ALWAYS THE ANSWER

Many people think the way to save gasoline is to drive as slowly as traffic will allow and depress the throttle as little as possible to maintain speed. We call that *featherfooting*, to make a comparision with the *leadfoot* who keeps the pedal mashed to the floor all day long.

Featherfooting is an oversimplification and is often wrong. It is definitely superior to leadfooting but not as good as intelligent driving for economy. Pretty soon we'll get into some cases and examples.

EVERY TIME YOU USE THE BRAKES, IT COSTS YOU MONEY!

When your car is rolling along at some speed, it has momentum, which means basically that even with no power it should coast for some distance before coming to a stop. You already know that it cost you some gasoline to get up to that speed, so you have an investment in momentum.

You can get back some of your investment by using the momentum. If you see a stop sign ahead, you should back off the throttle and let the momentum carry the car forward so you are getting some forward progress without using much gasoline.

On the other hand, suppose you are riding along at 30 MPH and suddenly use the brakes to bring the car to a stop as rapidly as possible. You didn't allow the momentum to pay back your investment, so you wasted it.

The only time a sudden stop through use of the brakes is justified is in an emergency or when traffic conditions require it.

If you could drive without ever using the brakes, you would get better mileage.

YOU CONTROL THE GAS WITH YOUR RIGHT FOOT

The foot pedal that you use to control the engine is commonly called the accelerator pedal or the gas pedal. What it actually does is open an air valve—called the throttle. This regulates the amount of air the engine can draw in through the carburetor. More air makes your engine run stronger and faster. The foot pedal should be called something like *air pedal* but isn't.

You probably already know that the carburetor mixes gasoline with the airflow on the way to the engine. The carburetor is supposed to mix in the correct amount of fuel for different operating conditions and it usually does. It functions automatically, depending on how much air is passing through, temperature and other things.

Some conditions cause the carburetor to dump in extra gas and you will naturally want to know what those conditions are so you can try to avoid them.

Cold Engine—A cold engine requires a lot of excess gasoline mixed with the air—otherwise it won't even run. The mechanical operation of furnishing the extra gasoline to make an extra rich mixture of fuel and air is called *choking.* In the old days, cars had a manual choke. You pulled out a knob on the dash board to start a cold engine and then tried to remember to push it back in when the engine warmed up and no longer required a rich mixture.

Modern automobiles have an automatic choke mechanism operated by a thermostat. The thermostat measures the temperature of the engine and, if it's cold the 'stat puts the choke into operation. When the engine warms up, the automatic choke is gradually turned off.

Because choking uses extra gas, and because an automatic choke is controlled by engine temperature, it makes sense to get the engine warmed up as quickly as possible when you first drive it in the morning, or after it has been parked a while. I'll tell you how in a minute.

Accelerator Pump—When you floorboard the gas pedal to make a quick getaway, a carburetor would fail temporarily to mix in enough gas. The problem is solved by a device called an accelerator pump.

A small reservoir of gasoline is held in a special section of the carburetor. A plunger lurks in this reservoir and whenever you cause it to, the plunger forces out the supply of gasoline, squirting it into the airstream flowing into the engine. This extra enrichment helps the engine accelerate strongly.

By a clever mechanism, the pump is only called into full action when you depress the gas pedal *quickly.* You can *ease it down* and the accelerator pump will squirt less gasoline. You will save gasoline each time you depress the gas pedal slowly rather than rapidly.

Power System—Inside a carburetor is a collection of small holes, called jets. They allow precise amounts of fuel to flow through, but all jets do not flow gasoline all the time.

When you open the throttle to accelerate or climb a hill the engine needs more fuel so additional jets or orifices are "turned on." These orifices are part of a system sometimes called the *power system* because it only works when you want lots of power.

If you could drive so you never turn on the power system in your carburetor, you would save a lot of gasoline.

BEST FUEL ECONOMY

Best fuel economy in normal driving usually happens when the engine is running under a light load. A light load means cruising on a level road at a constant speed so no power is being used in acceleration or climbing a hill.

Best mileage normally occurs with the throttle only partly open at a road speed of around 30 MPH. Wind resistance hasn't started to climb rapidly as it does at higher speeds, so the engine is just loafing along overcoming the rolling resistance.

SOME BASIC RULES FOR ECONOMY DRIVING

With all those facts in mind, here are the fundamentals of economy driving. These simple rules fit most situations and will make a noticeable improvement in your gas mileage—especially if you've been hotdogging it around.

1. Accelerate smoothly. Move the gas pedal slowly so you don't excite the accelerator pump in your carburetor.

2. Conserve momentum. Back off early when you intend to slow down or stop so you get your investment back out of the momentum. Use the brakes as little as possible.

3. Drive at a steady speed. If your speed fluctuates, you are accelerating unnecessarily some of the time, and acceleration costs money.

4. Keep your speed down as circumstances and your schedule will allow. Wind resistance increases rapidly as you go faster.

5. Drive smoothly. Look ahead, watch traffic and plan your moves. If you drive as these rules suggest, your passengers could have a tea party and never spill a drop. You can afford to furnish the crumpets.

PRACTICE

You may be surprised. Even these simple rules aren't all that easy to follow and we haven't really gotten into the nitty gritty yet. Consider it a challenge and you will probably find it becomes fun to do.

How much practice do you need and how much is enough? Well, in the days of the economy run many of the teams sent "advance men" over the route to be run. Weeks were spent compiling information which was put into a large looseleaf binder for each of the drivers. This binder contained every scrap of information to be found that would help the driver.

Obviously you won't have someone to prepare a loose-leaf binder of information for your driving—but the information contained in those binders is relevant because it shows what a driver should know to practice maximum gas mileage driving.

The terrain of each day's run was drawn on a chart which showed altitude and mileage. In other words, the driver could study the chart the night before the run and know exactly what sort of terrain he would encounter every mile of the way.

A lot of scheming and planning went into the economy-run contests. Notebooks like this were prepared in advance so the driver could take every possible advantage of terrain, route and traffic conditions. You probably won't ever go to this extreme, but you can use the same basic principles in your driving.

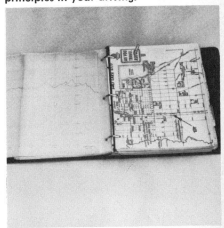

15

Even the way you stop has a big influence on fuel economy. Get off the gas early and use up the vehicle momentum in coasting—rather than heating up the brakes and wearing them out.

Truck and bus drivers are masters at passing without altering speed. They look ahead, anticipate traffic and make smooth moves, as this view through a streaked windshield shows.

During the run a navigator would be sitting in the rear seat with the chart, passing on the information to the driver.

"You're coming to a long grade. It is six-miles long and goes from 1000-feet elevation to 2000 feet. The grade drops off sharply when you see a 'Protect National Forest' sign on the right."

Neat huh? That's just the beginning. Every stop sign, stop light and railroad crossing was documented. The driver not only knew how long the light would stay red, but how it was tripped—automatically or by vehicle and if there was a pedestrian trip.

Radio stations in each area were noted in the book. Town, frequency, call letters and the time of day weather reports would be issued were listed so the driver could call ahead for information during a lunch break. Would he be encountering gusty winds during the afternoon? From what direction? Would that help or hurt?

With all of this information plus practice in controlling the cars precisely, it was possible to crest a hill at perfect speed for maximum mileage benefits, back off the throttle for a downhill run into a town and with the help of a sharp navigator and a stop watch, sail through 133 traffic lights without making one stop. That actual number of lights was noted in one town in one of the log books. Needless to say, you don't pull this off without some practice— and this is the kind of practice that leads to great gas mileage. While this might not be practical for you, it does give some insight on how the economy wars were won.

The battle is not yet over and practice makes the difference. At the Union 76 Economy Tests in 1975, we had to drive according to a precise plan: Accelerate and brake at certain rates and drive at prescribed speeds—in addition to meeting a time schedule. Some drivers did not find or take time to practice. These drivers did poorly in gas mileage and in some cases failed to qualify their cars.

Practice may not make perfect—but boy it sure can help your gas mileage!

ECONOMY DRIVING TECHNIQUES

Now let's get into the driving techniques you should use. If you really get serious about developing the driving skills for increased mileage—and are willing to apply them each time you drive—they will give you free miles every day.

APPROACHING STOP SIGNS

When you are in fairly open country and traffic is not particularly heavy, start slowing 1/2 mile from a red light. This doesn't mean you should take your foot off the throttle when you see a red light 1/2 mile away. If you do, you'll find yourself parked on the road somewhere before the stop. Before that happens you will realize the slowdown is too rapid and consequently you will accelerate about 1/4 mile away from the stop sign. This is exactly the type of driving we're trying to avoid so let's go through it step by step.

The ultimate in economy when approaching a stop can be attained by learning to lift the throttle gradually. The moment you spot the stop sign or signal light, select a spot halfway between you and the stop and at the same time start to ease off the throttle. Plan on dropping 2 to 5 MPH by the time you hit your selected halfway point. When the distance to the sign is approximately 1/4 mile—back off the throttle. Now you can stop with easy braking. Although this sounds simple, some practice is needed to work this technique to its fullest.

Right here is where you start planning your move to get around the car ahead. If you wait until you are right behind, you will probably have to slow down a lot, then accelerate a lot.

Generally highway signs are placed quite consistently at a given distance from the intersection. Perhaps that could be your guideline—the intersection sign is the warning of an upcoming stop sign. If your depth perception is reasonably good, just pick a spot you think is halfway between the point where you are and where you may end up stopping and perform the deceleration as outlined.

Surely you wonder about coasting out of gear. It is illegal to drop the vehicle out of gear and coast and it is very dangerous to drop a vehicle out of gear and turn off the ignition. Engine-driven power accessories such as brakes and steering quit helping you. On some cars and trucks with anti-theft locking steering columns, the steering wheel locks when the ignition is off! The savings you get by coasting are not worth the trouble and risk.

In city driving, you usually don't have the luxury of taking a half-mile or quarter-mile to get slowed down and stopped. However the basic idea is the same. Back off as early as traffic will permit, use up the momentum to carry you to the stop sign, use the brakes as little as possible.

PASSING MANEUVERS

If you are on the open highway and overtake another vehicle, it is poor economy to slow to the speed of the other vehicle, follow it for a while, and then have to step on the gas again to get up to a satisfactory passing speed. It is much wiser and more economical to plan your passing maneuver as soon as you see that you are overtaking. Observe what is beyond the car you're going to pass. If the road is clear, continue without changing your speed and go around the other car. If, during the passing maneuver, a car appears in the oncoming lane it is more economical to accelerate slightly to make a nice clean pass rather than slow down abruptly, get back in line and wait for another opportunity to pass. This is a case when leadfooting briefly is the answer.

How is it possible to violate the smoothness rule we keep laying down and still save gas? Let's say you are overtaking a large slow truck on a fairly level road. Getting into the throttle and passing the truck *now* could prevent you from having to stay behind the truck during a long, uphill grade which is just around the next bend. If you know that grade is there and you know there will be limited, if any, opportunity to pass for several miles then get around the truck and set your own speed up the grade.

If you are really concerned about better gas mileage you'll cringe at this sort of situation. In an economy run a driver will paint the air blue for an hour or so—cursing his luck at having to face such a situation and having to choose the lesser of two evils!

Always think about the safety aspect of any driving maneuver. Any time you drive erratically or make unnecessary moves—whether it be steering, braking or accelerating—it constitutes bad safety habits and is very poor fuel economy practice.

When faced with a situation where you can't pass because of road conditions or oncoming traffic—lay back a little and occasionally pull slightly to the left to observe what's ahead of the vehicle you want to pass. When it's obvious you're going to get to pass, start to accelerate while you're still directly behind the car. This *requires* the extra space we mentioned earlier.

This gentle acceleration before getting out fully into the passing lane gives you passing speed and reduces the length of time you are exposed to the additional risk of passing. Truck drivers make an art of getting up speed and then swinging around the vehicle ahead at just the right moment. All drivers should practice this technique conscientiously. It is the safe and economical way to pass.

In the city, the passing problem is much more complicated as you already know. You have to think about passing individual cars and also about getting by groups of cars and the effect of traffic patterns.

For example, you may be on a street which allows left turns from the inside lane rather than providing a special lane for drivers waiting to make a left turn. You learn to watch ahead for drivers planning a left turn and if you can safely do it, change to the right lane so they won't cause you to slow down or stop. Sometimes drivers up ahead will actually operate their turn-signal lights to give you a clue about what they are planning to do.

What you would like to accomplish is to make your entire trip through city traffic at a constant speed without ever

Because this road curves to the left, scanning ahead requires you to glance to the left. Suppose you only look as far as line C. You could be trapped into passing the car ahead on the right side and then getting caught behind that camper in the right lane while the car you are going around shuts you off at the pass.
If you look as far as line B, you'll see that the car towing a trailer is controlling traffic behind it and you'll decide to stay where you are and watch what develops.
If you look as far as line A, you'll realize that nothing is going to unscramble this cluster of vehicles for a while yet, so you will relax and just cruise with the traffic. By the time you are this close to the car ahead, you should already have made that decision.

touching the brakes. You can rarely do that but you can try.

Suppose you see a school zone ahead with a sign holding traffic to 15 MPH, or you remember that one is ahead. If there are no other cars between you and the school zone, you will naturally ease off the throttle and collect some of your momentum investment while slowing down. If two or three cars are ahead of you, what's going to happen? Usually each driver is surprised by the fact that the car ahead of him slowed down. He waits too long and has to apply his brakes to keep from crashing into the car ahead. When there are several cars doing that, the last in line is likely to come to a complete stop or a slow crawl.

If you are tucked in close behind the last car of the parade, you will have to slow to zero or one MPH. If you'd planned ahead you might only have to slow to 15 MPH. Allow for the traffic congestion and ease off your throttle even sooner. Braking costs money!

TAILGATING

Tailgating means driving too close to the car ahead.

If you have decided to drive for economy, you have made the *basic decision* that you are going to manage your own driving operation and not let other people do it for you. But if you are close to another vehicle, you are being controlled by the other driver. If he slows down because of poor planning or not looking ahead, you absolutely have to slow down also or bend his bumper.

It's a sad fact that nearly all drivers seem to have an urgent desire to tailgate somebody. If you leave a reasonable amount of room between you and the car ahead, it'll stimulate the guy behind you into temporary insanity. Maybe permanent.

The fellow behind will honk, yell, pound on his steering wheel and do everything possible to urge you to close up the gap. If he gets the slightest chance, he will dash around you and snuggle up to the car ahead.

You have to make a decision about situations like that so you don't wind up being controlled by other drivers. It's easy when you realize that the only way to keep those maniacs from tailgating the car ahead of you *is to do it yourself* so they can't get in.

That's plain ridiculous. You are not going to tailgate to prevent somebody else from doing it. Are you?

MOUNTAINS, GRADES AND HILLS

When it is obvious you're climbing a long grade, do not accelerate or make any significant change in your driving speed because of the grade. Surrounding terrain sometimes makes it easy to be deceived as to whether the road is flat or a slight grade but your speedometer will tell you.

Long Upgrades—Let's consider a one- to five-mile uphill grade that is just a long gentle climb. The additional load on the car will take a little more throttle to hold the constant speed you've been running on level ground. Do just that.

If your cruising speed has been 55, maintain that speed. As you feel and see the speed falling off one MPH or so, start applying gentle pressure on the accelerator. Use just enough additional power to maintain the speed you've been running. There is no point in holding a fixed throttle opening because this is going to drop the speed and put you over the top at a very low speed. Accept the fact the hill is there and you have several thousand pounds of car which has to be lifted over the hill. You might just as well do it at your level-road running speed.

In summary, when approaching long grades anticipate when the car will reach the grade and will start to slow up. Just as speed starts to drop, ease into the throttle slightly—maintain normal driving speed right up to the crest of the hill. As you approach the top, learn where to back off the throttle so you don't gain speed while crossing over the top or starting down the other side.

One of the most popular forms of freeway accident is caused by tailgating at 60 MPH. The driver in that Volkswagen can't even see traffic flow beyond the truck up ahead. When traffic comes to a sudden stop, the Volks may require stopping assistance from the car in front.

In this mess, the chap in the station wagon with box on top is the only one showing good sense. The car behind is probably honking and yelling. Photo courtesy of Arizona Department of Transportation.

When climbing a long grade, the general rule is to maintain your normal highway speed by feeding a little more gas. An exception is underpowered cars—discussed later in the book.

Get back some of the gasoline energy used to climb up the hill by easing off the throttle while going down the other side.

Long Downgrades—If you follow the rule of maintaining your normal cruising speed, you will automatically do the right things. Just back off the throttle and get back part of the extra gas you used to climb to the top of the hill in the first place.

Steep Downhills—Stay completely off the throttle while going down. Don't downshift a manual or automatic transmission to use the engine for braking unless it is necessary to control speed. If you can control speed with one or two gentle applications of the brakes while going down, that's better than downshifting and causing the engine to run faster all the way. However, don't use the brakes for any long period of time on any downhill run—they may heat up and the braking ability will fade.

You Are Stopped at the Bottom of a Hill—If you are stopped by a signal or stop sign just at the bottom of a hill, or partway up the hill, you face the double penalty of having to accelerate and having to do it while going uphill.

In city driving, traffic may make it worse so if possible choose a route to avoid this situation.

On the highway, with an automatic transmission, accelerate briskly to speed and go on up the hill. With a manual, accelerate and shift quickly between gears to avoid losing momentum at the shift points. On reaching the gear which will climb the hill without laboring or lugging, maintain speed on the rest of the climb.

Steep Rolling Hills—In hilly terrain, try to drive at a constant speed. Work the throttle gently to maintain speed both uphill and down.

GEORGE OBSERVER SLEPT HERE

In the early days of the economy runs, cross-country trips were used to test economy of the cars. A car would have a driver, a co-driver (who normally would feed information to the driver on the route and traffic situations) and the observer.

The observer rode in the front seat and it was his job to make sure the driver played by the rules—that he stayed on course, did not violate traffic laws and didn't kick the car out of gear and coast down long hills.

The observers suffered from human frailities with which we are all familiar. One observer admitted to his driver that he had fallen asleep on two previous runs and that if he were caught one more time he would be fired. The driver began to devise a plan. Knowing of a very long downgrade to be encountered later in the day, the driver and co-driver ceased all conversation for fifty or so miles before the long hill. The heater was turned up—and then as the observer began to nod the heat was turned up further. With the interior of the car being toasty warm and no words between driver and co-driver the observer's nods turned to a deep-breathing slump right on schedule. Once on the downgrade, the driver ever so smoothly eased the car out of gear and into neutral. The long free coast was on and time was being made; that bus was rolling.

At a later point during this joy ride the observer awoke with a start. The driver just as quickly tried to get the car back into gear. The sudden move found the driver putting the car into low gear instead of high. Engine speed screamed. The vehicle was pulled almost to a stop. The observer rummaged for pad and pencil to write a protest. But the driver made his point and made it well. If the violation for coasting got turned in, so would the complaint from the driver that the observer fell asleep during the run. The argument ended in a draw. After the unplanned shift into low gear, no one will ever know if the long downhill coast saved gas or if the shift cost far more gas than was saved.

There are very few tricks that weren't tried by the competing teams in the early economy events. Since then the supervision and attitudes have become more sincere and contests today are more an evaluation of vehicle mileage than team ingenuity.

HEAVY TRAFFIC

City freeways and major streets are often plagued with stop-and-go driving. When faced with bottled-up lanes and slow traffic, take the approach you do in any kind of driving—be reasonably steady and look ahead. It is very important to be aware of what's around you from the standpoint of safety as it is for improving economy.

Changing lanes suddenly is against the law. Constantly scan ahead and to the side. This provides quick observation of an

Commuters and through traffic merging on the expressway make more stopping than going. Try to move forward several car lengths at a time, rather than starting and stopping constantly to keep zero distance between bumpers. With an automatic transmission in start-stop traffic, you can often just let the car creep along slowly. Arizona Department of Transportation photo.

upcoming bottleneck or a slow-moving vehicle. You paid for the mirrors on your car—use them. Be aware of what's going on 360 degrees around your immediate area or the hope of driving smoothly and economically becomes a pipe dream. If you practice, you'll find it's easy to scan a quarter- or even a half-mile ahead of you all the time.

Plan your moves and don't be frustrated by the traffic problems. Do the best you can but face the plain fact that economy driving is difficult or impossible in a traffic jam. There are a few things you can do.

When you are bottled-up in traffic and it's stop and go, stop and go, the fewer times you have to use the accelerator the better off you'll be—especially with an automatic transmission. If traffic is moving at all, just try to maintain a very light throttle and let the car drift along with as little engine assistance as possible. If it's a stop-and-go situation, most drivers will want to move a car length or even less every time they can. Try to delay moving forward until you can make your moves 40 to 50 feet at a time. In the meantime sit there and idle. Idling an engine takes very little fuel. If you idle long enough of course you can use up an appreciable amount of fuel. But 10 to 15 second idles are very insignificant in the total amount of fuel you will use in driving for a half hour.

Rest assured that in a stop-and-go situation an economy-run driver wouldn't get in any hurry to move forward even with a couple of horns blaring away at the rear. In fact if you really want to get serious about it, disregard the horns and shouts until a larger protester starts to get out of his car! Be very light on the throttle when you do have to move the car from a stop. Move as seldom as possible; plan your moves.

Sometimes an amazing thing happens. People behind you get the idea or at least discover that moving longer distances each time is less frustrating. A whole line of drivers will suddenly

Let the other guy race up the intersection and stop at the red light. By planning ahead and adjusting your speed, you can get there just as the light changes to green and roll through without using your brakes at all.

become relaxed, accept the fact that they can't push the line ahead to make it go faster, and on rare occasions they have even been observed showing courtesy to one another.

In the days of the economy runs, contestants were fully aware that poor mileage figures were generated in metropolitan areas. As a result drivers spent the major part of their practice time working in traffic—working at being smoother blending from one lane to the next and timing traffic lights. The route was set and could not be varied. In this respect the average driver has it much better because you are not required to drive along a specific route and sometimes you can even choose the time-of-day for your trip.

TRAFFIC LIGHTS COST DOUBLE

If you have to stop at a red light, chances are traffic conditions wouldn't allow you to coast down and you had to murder your momentum by using the brakes.

Then when the light changes to green, you have to rebuild the momentum by accelerating up to your cruising speed again. You waste gasoline *twice*.

What you want to do, obviously, is *sail* through the intersection without even slowing down. The forces of law and order prefer you do that only when the light is green.

So, you learn to read the lights. Watch ahead and make advance alterations in your speed so you hit the light while it is green. There will usually be some joker behind you who wants to hurry up to the intersection so he can stop and wait for the light to change. The facts are, if he follows your pace he will not only save gasoline but will probably get down the road a little faster anyway. How many times have you passed cars that were stopped for a red light by getting there just as it turned to green? So don't worry about the joker behind you.

Some streets have traffic lights set so they will allow travel at a certain speed without hitting a single red light. It's more common than people think because it usually isn't advertised. Also, because some lights are operated by a clockwork mechanism somewhere they have to be on *some schedule* and if you look for it you may find a way of driving that gets you mostly greens. This can involve going a little slower from Main to Broadway and then speeding up some from Broadway to Western, or some such.

Occasionally you will get into a funny situation. Suppose you are in the curb lane and a hotdog is in the inside lane. The light changes and you both motor away—you conservatively and he with a blast. At the next signal he gets caught by the red and stops. You have it timed right so the signal changes as you arrive and you roll on through. In about three seconds he goes fogging by so he can stop again at the next signal.

You can pass the same car ten times that way and it only serves to arouse the competitive instincts of the other driver. He burns more rubber to get ahead of you each time and then waits longer for the signals to change.

You can actually enjoy little traffic games like that just as much as a hairy drag race at each signal. Keep your cool and save your bucks.

GEAR SELECTION

Cars with automatic transmissions are pretty poor on economy up to about 30 MPH. This varies from one car to another—but generally getting up to 30-50 MPH is very inefficient due to the slippage in the converter.

There's not much you can do about it except avoid frequent accelerations. A manual transmission is essentialy a solid coupling each time the clutch is engaged with the car in gear, so there is no slippage.

If you have a car with manual transmission you may have thought about what gear to use in a traffic situation that is stop and go. Use the highest gear you can without lugging the engine which may cause U-joint clunking. Keep just enough RPM so the engine isn't chugging along and yet isn't turning unnecessary RPM which will be obvious from the sound.

The engine, transmission and rear axle wear faster as a result of erratic driving. Steady driving will make the entire drive train last a great deal longer. This is the indirect saving; not the primary thought, but a saving nonetheless.

So far as we're concerned there's always an energy crunch. Our energy is what it takes for us to buy things, repair things, keep up with inflation and have some entertainment. Sure it's a type of energy, but as long as we're talking about saving, let's save in every way we can.

WIND, WIND, GO AWAY....

In most parts of the country there's a good chance you're going to encounter wind—sometimes very gusty, high winds. Of course, if you are traveling with the wind it can help tremendously in increasing gas mileage. But if the winds are hitting the vehicle at the front or side, this is damaging to economy. You would think a head wind would be much more damaging to gas mileage than a side wind. Our testing indicates this just ain't so! We found there may be only a small difference in the effect of head winds and side winds.

So whenever either adverse condition exists, you should slow down a bit, because the harder you fight the wind and the harder you push the throttle to maintain speed, the more fuel you're wasting. Slow down a little—somewhere along the way you can count on picking up a tail wind or a period of calm and gaining back the time you lost. But you'll never gain back the good miles per gallon numbers you lose when fighting the wind.

Here's an example. If you are driving 50 MPH into a 20 MPH wind, that's the same as driving 70 MPH on a calm day with no wind. As you saw in the earlier table, wind resistance at 70 is nearly double that at 50 MPH. The fuel penalty to maintain 50 MPH will be very high. Slow down a little and plan to make up time later.

If you are driving 50 MPH with a 20 MPH tail wind, you are effectively moving 30 MPH as far as wind resistance is concerned. With a tail wind or no wind at all you can hotdog it a little to make up time without a pronounced effect on fuel economy.

I mentioned earlier that economy-run contestants are given phone numbers of the local aviation weather service at points along the route. Let's say the run is from west to east and the forecast is for surface winds toward the west in the morning, reversing in the afternoon to a wind that blows toward the east.

The driver knows he is facing a head wind in the morning and will have a tail wind in the afternoon. He will plan to run a little slower in the AM and a little faster in the PM, to maintain the desired overall average speed.

Incidentally, reversing wind directions are typical of storm fronts. The wind will blow one direction at the beginning of a storm and the other at the end, when the front has moved farther along. This can help you on a long trip when you want to get some gas mileage you can brag about.

It's easy to see that head winds and tail winds either oppose progress of the vehicle or else get behind and push.

Side winds have a little different effect. For one thing, even casual streamlining of the body shape is based on the assumption that the wind is coming from the front. If the wind comes at an angle, the car effectively presents a larger area and wind resistance is higher.

Another effect of side winds is to increase the rolling resistance of the tires. To hold a straight course with a side wind, you will normally steer into the wind a little bit and the tires have to exert a side force against the push of the wind. This causes the tire to distort in a different way where it is in contact with the pavement. The extra distortion takes extra energy and more gasoline from the tank.

Either way—head wind or side wind— you are better off yielding a little than fighting it all the way.

In open country, wind is common and can affect economy. If it is from behind, it helps. Head winds and side winds reduce economy but you can help a little by slowing down and not fighting it so hard.

ECONOMY-RUN TRICKS

I don't think the average driver should spend a lifetime driving like an economy-run driver. Many of the economy-run drivers also double as racing drivers blessed with lightning reflexes, sharp eyesight and a very good "feel" for a car. I point this out in preface to telling about one driving "technique" that panicked a few observers who were riding with contestants.

If the economy-run car was being met by a large truck on a two-lane road, the economy driver would move to the far right of his lane, then as the truck got very near—30 feet or so—the economy-run driver would pull back to the left and come quite close to the back end of the truck. A really good driver can come very close—certainly close enough to scare the wits out of an observer in the back seat.

The reason for the technique is this: A large truck disturbs a lot of air and creates turbulence which is quite harmful to the economy of an on-coming car. If the car stays to the far right—away from the truck—it will be in the turbulence for a longer period of time, but by "hitting the air" just as it leaves the truck and being

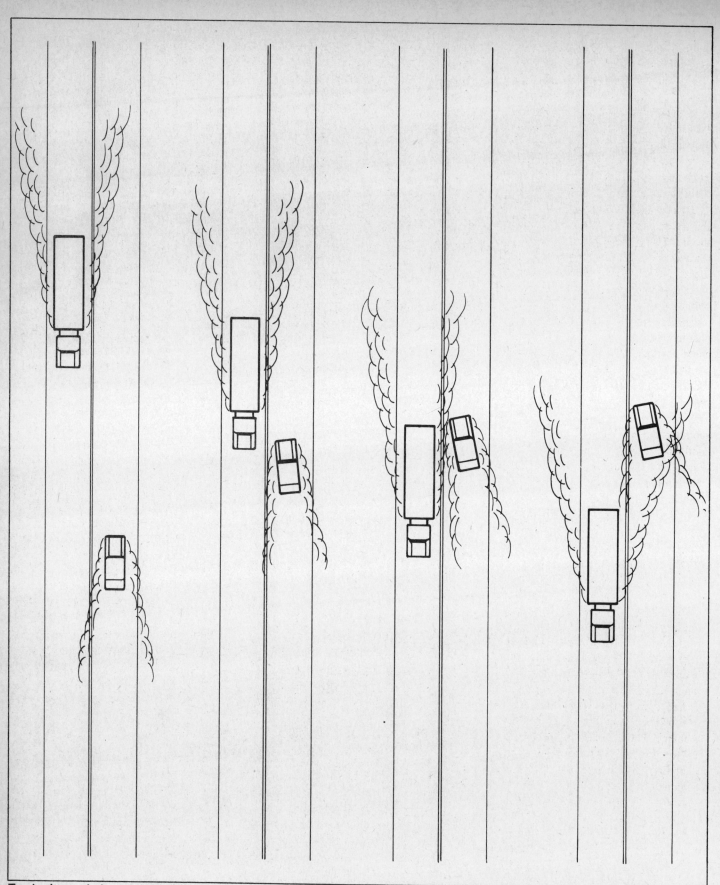

Trucks drag turbulent air behind them which buffets a car coming from the opposite direction and reduces gas mileage slightly. Economy-run drivers cut in close to get through the turbulence in the shortest time. The improvement in fuel economy is not worth the risk for the average driver.

as close as possible to the truck the length of time the car is in the turbulence is reduced.

An economy-run driver would also attempt to get a "drag" or tow from a large truck by using the truck to block the wind. If the economy driver came up behind a large truck traveling close to the correct speed, he would stay 20 to 30 feet behind the truck for as long as he could get away with it.

This has been a racer's trick for years and the economy-run observers know all about it. The observer would start to protest and the driver would start to argue—first claiming the truck was going the exact speed he wanted to travel, then switching his argument to point out that it was now dangerous to pass due to oncoming traffic. The longer the driver could argue with the observer, the longer the tow. The driver backed off or passed the truck only when the observer pulled out his pen to write a violation.

Some of the shenanigans at the economy runs were classic examples of good old competitive "psyching." One year a factory team hired nothing but state troopers to drive their cars in a run held at a race track. Starts and stops were part of this particular economy schedule. The factory which hired the troopers played their angle for all it was worth. Drivers were trained, practice sessions held, press release photos taken and interviews granted. All was ready.

While this was going on a competing team learned through practice that all of their full size, stick-shift cars would get better mileage if the standing starts and the one-two shift were made briskly enough to burn rubber. The troopers watched in amazement as their competitors took the trophies for that day. Grabbing rubber on economy schedules and "winning!" Too much to take—their instructors hadn't mentioned this technique.

The following day when the compact cars ran the course the trooper drivers drove their teams cars with vigor—burning rubber and wisps of tire smoke were evident on all starts and shifts. The team responsible for initiating all of this motored sedately away from the start line and eased through the shift pattern, using the accelerator as little as possible. Needless to say, the compacts driven by the troopers lost by a country mile because the monkey see, monkey do tactic was only intended for one particular line of full size vehicles—not compacts.

There is a moral here. Just because a gas-saving tactic works for one driver with one car doesn't mean it will work for you and your car. Do your own research.

AXLE RATIO

In a conventional car, the engine and transmission are up front, connected to the rear axle by a drive shaft that runs under the floor. Through some special gears in the rear-axle housing—called the *differential*—the drive shaft causes the rear axles and rear wheels to rotate.

The axle ratio is a number which tells you how many times the drive shaft must rotate to cause the rear wheels to rotate one time. If the ratio is 3.70, for instance, the drive shaft must rotate 3.7 times to cause one complete revolution of the rear wheels.

A lot of slang expressions used by mechanics and car nuts only confuse the casual motorist. They speak of "tall" and "short" rear axles, also "long-legged" and a lot of other names. "Gearing down" usually means changing the ratio to a higher number and "higher gearing" usually means changing the ratio to a lower number.

If you just find out what the axle ratio is, by number, and use the definition given above, you will avoid confusion.

FEATHERFOOTING IS NOT ALWAYS THE ANSWER

Very often in newspapers articles and magazines you will read all about avoiding jackrabbit starts and high speed. It ain't that simple: Depending on the weight of the car, the torque of the engine, the axle ratio, whether you are headed uphill or down, you might get better mileage by accelerating from a stop rather rapidly—just short of getting into the power system of the carburetor—and then dropping it into high gear or letting the automatic transmission shift for you.

As an example, here's some actual data taken during pre-run testing for the 1964 NASCAR Pure Oil Trials.

A one-mile section of mountain road, climbing and then descending with 3 to 4 percent grades was used for testing. The object was to determine the best driving method for maximum mileage and still maintain an overall speed of close to 40 MPH. Sounds pretty simple doesn't it? The accompanying data should show you that the admonishment to featherfoot is often simplistic and incorrect.

This data was gathered by testing a 327 cubic-inch V8 Chevrolet with an automatic transmission and 3.55 rear axle ratio. This isn't somebody's opinion, these are real numbers from a real automobile on a real road. It's worth spending some time looking at the data carefully.

Four tests were made with different uphill and downhill speeds so the overall average was about 40 MPH. Column 1 shows the up and down speeds. Column 3 shows the uphill and downhill speeds combined resulted in close to 40 MPH as the average speed.

Column 2 shows MPG figures at different uphill and downhill speeds. You can see the slowest uphill speed of 25 MPH gave best fuel economy for the uphill part. The slowest downhill speed

Four tests, driving up a hill and down the other side, show that featherfooting is not always the answer. Tests are described in text.

Test	1 Maximum Indicated MPH		2 Actual MPG		3 Average MPH Based on time and distance	4 Average MPG
	Up	Down	Up	Down	Up & Down	
A	25.0	55.0	12.38	58.3	40.03	35.0
B	30.0	50.0	12.33	59.2	39.24	36.0
C	35.0	47.0	11.32	63.1	40.60	37.0
D	40.0	46.0	10.5	72.8	43.01	41.0

of 46 MPH was best in that department. Don't form any conclusions yet.

Column 4 shows the overall average miles-per-gallon, combining both the uphill and downhill sections. Test D gave the best overall result.

Now let's look at some details. In Test A, climbing the hill at 25 MPH gave better gas mileage than going up at higher speeds. Descending at 55 MPH gave the poorest of any test. Why?

If the car climbed the hill at 25 MPH and then went down the other side at 55 MPH, it had to accelerate at the top or going down. That used gas and got poor economy on the downhill run.

Tests B and C show about the same thing. The downhill speed was quite a bit faster and therefore the driver was feeding some gas.

In test D, the car went up at 40 and came down at 46. Because it was downhill, the increase in speed was due to gravity and not to gasoline.

Among all the different tests, D is best overall economy at 41 MPG. It is also the one with nearly constant speed uphill and down. It looks like the driver intended to hold a steady 40 but the downhill run speeded the vehicle just a little. Nobody would put on the brakes if he could avoid it.

Notice that featherfooting it up the hill got less economy than what we advised you to do right at the beginning of this chapter. Recognize that you have to drag these older cars up the hill at normal highway speeds for best economy.

You may be wondering if going up according to Test A but coming down according to Test D wouldn't give the best overall economy regardless of the speed average. We didn't test that way but the results would not be as good as one might think. Because, to do that, you would have to speed up from 25 to 46 MPH on the downhill run and that requires feeding fuel to the engine.

The real reason a downhill run at 46 MPH is better than the others is not because it was at 46 MPH or any other particular speed. It's just because the driver got off the throttle and used gravity to drive down the hill. The best overall gas mileage would probably be going up at 25 and coming down under gravity at 28 or whatever it turned out to be. But you can't do that and maintain an average of 40 MPH so we didn't test that either. Also, that's pretty slow out on the highway and most people would pay a little more in fuel to go at a better speed.

In this example, going up the hill slowly while using your tender touch on the throttle is not the best way to do it if you are after some average speed during the entire operation.

Some current production automobiles often have low numerical axles: 2.56, 2.76, etc. With these gears the best way to get economy is to use momentum as much as possible to help climb the hill and allow the car to slow down while climbing. Pick up momentum again while going downhill. Small cars today still have high numerical axles such as 3.07, 3.36, 3.73, 4.11, etc. These can be driven briskly in hilly situations without hurting economy.

In the following test a 327 cubic-inch Nova with an automatic transmission and an axle ratio of 3.2 was accelerated *from a dead stop* to 30 MPH with the vacuum gage kept at a predetermined set value by varying the amount of throttle. The time it took to get to 30 miles an hour, the amount of time to travel a 1/2 mile distance and the amount of fuel used in this distance were recorded.

Let's say this test represents four different drivers in identical cars, all sitting at a red light. When the light changes, each is going to accelerate up to 30 MPH, which is the speed limit, and then hold 30 MPH indefinitely.

Driver A is super conservative because he has read in the newspaper that featherfooting is the way to go. He takes 38 seconds to reach the speed limit, holding a reading of 15 on his vacuum gage.

Driver B holds a vacuum reading of 10, gets up to 30 MPH in less than half the time, but still is driving according to conventional conservatism.

Driver D is the stop-light drag artist we met earlier. He zooms up to 30 MPH in only 9.5 seconds and then holds that speed because he thinks the law is watching.

Driver C is you. You use half-throttle or a little more because you know that engines are inefficient at very low RPM and automatic transmissions waste gas below about 30 MPH. So you didn't spend a lot of time in the zone of inefficient operation. You got it up to speed without turning on the power system in the carburetor, but you did it with reasonable zip. Ten seconds to go from zero to 30 MPH is not exactly a world's record for sizzling automotive performance. But quick enough so people won't honk at you from behind.

Notice that driver A took nearly four times as long to accelerate up to the speed limit compared to driver D. Now look at column 3, the time to travel a half mile. The hotdog was ahead by only about 6 seconds. Proving once again that the stop-light drag artist gains mainly attention and tickets.

Driver C used the smallest amount of fuel to travel a half mile. You rascal!

This data represents four different drivers in identical cars, all accelerating from stop to 30 MPH. The vacuum readings in column 1 were taken with a test instrument and basically show how much the throttle was opened by the driver's right foot. A high vacuum reading such as 15 means the throttle was nearly closed. Driver A, the featherfoot, and Driver D, the leadfoot, both used the same amount of gas to travel 1/2 mile. Which of these four drivers is you?

Drivers	1 Vacuum held during acceleration standing start. (inches Hg)	2 Time to 30 MPH	3 Time to travel 1/2 mile	4 Fuel used to travel 1/2 mile
A	15	38 sec.	69.8 sec.	77cc
B	10	16 sec.	67.1 sec.	80cc
C	5	10 sec.	64.4 sec.	63cc
D	Wide Open Throttle	9.5 sec.	63.6 sec.	77cc

SHORT TRIPS ARE WORSE THAN LONG ONES

We have been brought up in an age where the family car is regarded as a convenience akin to the light switch on the wall. Use it. And like the light switch, use of the family car has drifted toward the "abuse" side of the ledger.

A trip to the grocery store for a loaf of bread now and a trip an hour later to the same shopping center for some needle and thread. An hour later off to the same grocery store to pick up popcorn and beer for the night's viewing session.

Because the engine in the car never fully warms up during these short trips, the choke is operating to a certain extent until you pull back in the driveway. As a result gas mileage is often less than half of what could be achieved on a longer trip when the engine has time to warm up.

The message is clear: Combine the family errands and come up with a one-hour trip in the car instead of five or six completely separate trips to the same area in the same day.

WHAT TO DO ABOUT COLD STARTS

In the days of the economy runs drivers had to go to their vehicles in the morning, start up and move out of the impound area on a given signal. No time was allowed for warmup.

Hundreds of pre-run tests were conducted with fuel meters to determine how best to handle the situation. Should the driver pull over and let the engine warm fully, then start the day's run? Should he creep along slowly and let the engine warm up?

The tests showed that with the automatic choke on the engine was doomed to consume a given amount of fuel per minute. That being the case, the plan was to get the car up to speed and out on the highway as soon as possible.

Racing an engine or "gunning" it while cold or attempting to accelerate very rapidly when it is first fired can damage the engine, but this danger passes after 30 seconds or so when oil is in full circulation.

You can apply this economy technique to your own everyday driving. Firing up and backing out of your garage or parking space allows enough time to get the oil circulating and distributed throughout the engine. This will provide adequate protec-

On a cold start, you'll serve both your wallet and your engine better if you just warm up long enough to get the oil flowing and the engine willing to move the car. Then drive conservatively for a few miles without lugging or racing the engine, while it is getting the rest of the chill out of its bones. You will get it off automatic choke sooner by driving than by letting it sit there and idle.

tion against wear if you drive conservatively for the first few miles. Pull away with gentle acceleration and drive at a moderate speed. That way you get some benefit from the gasoline you are burning while the engine warms up.

Another tiny tip from the experts: If you park outdoors, try to park so the first rays of the morning sun will fall on the hood and give the engine a bit of initial warmup. Why not—it's free!

TARDINESS

Keeping yourself on time or, better yet, ahead of time by starting earlier is good practice. When you find yourself behind schedule but feel morally obligated to be on time, you will sacrifice economy for a little bit of hurry-up. Sometimes people also sacrifice safety and pay for their error.

Get up in the morning, take a good look at the clock—plan your schedule. Have your cup of coffee or breakfast. Do it a little early. Rather than read the paper and then go to work, perhaps you should put the paper under your arm, scoot off to work, be there a few minutes early and take those few minutes to read the paper. Being late never buys economy.

IMPROVE YOUR ROUTE

Give serious thought to taking a different route to work or wherever you travel regularly. Often your regular route is not the best for economy. It may be the shortest—but still not the most economical. With a little bit of thought and experimenting, you may find it pays to go a different way and avoid a couple of stop signs, a red light or two, or maybe a clogged intersection. This might not be obvious at first, so work at it.

Check a map for alternate routes if none come readily to mind. Run each new route for a couple of tankfuls and compare this to a similar test conducted over your present route. If your driving is routine and fairly consistent from one day to another, then perhaps a better route will net you a saving of up to a mile per gallon.

Consider the condition of the road. A rough road will increase rolling resistance, shorten the life of the front suspension components and generally loosen up the rest of the vehicle. Constant jiggling on rough roads may bounce the valve that controls gasoline flow into the carburetor and cause a high fuel level. This means an over-rich mixture going into the engine.

A smooth road means economy—in more ways than one. Bypass that rough section and find another route.

USING THE TRANSMISSION

Whether you have automatic or manual transmission, as a general rule you want to be in the highest gear the engine can pull without lugging or bogging down. The sound of the engine should tell you it is operating near the middle of its speed range, neither chugging nor racing.

Manual Shift—If you have a stick shift, you have complete control which sometimes means you are allowed to do it wrong as much as you please. It's better to do it right.

Don't wind the engine out in each gear. Take the engine speed just high enough so when you shift into the next higher gear the engine will accept the load without seeming to bog down or struggle with it. Don't speed up the engine unnecessarily while the clutch is disengaged.

Shift up to the higher gears as soon as you can. If you get slowed down by traffic, or get on a hill so the engine does not sound comfortable in the gear you have chosen, shift down to the next lower gear until traffic or road conditions change.

Automatic—Automatic transmissions are controlled both by what the engine and car are doing and what you are doing with your right foot. If you keep the accelerator pedal on the floor, that smart transmission will know you want strong acceleration. It will stay in a lower gear until the engine runs up to a high RPM—then it will shift to the next higher gear and do it all over again.

That's exactly what you don't want for economy. You want the same operating conditions described above for the manual-shift situation.

Normally the transmission will shift down whenever the engine is about to lug and it is about as good as you are in deciding when that needs to be done.

The transmission may not be as interested in economy as you are and it may not shift up as soon as it could. You can help it a little. After you have driven a car for a while and have the "feel" of it, you begin to sense when the transmission could shift up if it wanted to. You can

On your regular trips such as to and from work, experiment to find the best route. Sometimes a combination of side streets and main roads will save you both time and money.

make it shift at that time by backing off slightly on the throttle.

Practice this technique and you will find that you can control the upshifts of an automatic nearly as well as your neighbor with the stick shift. Then use your talent to shift up a little early and get the advantage of the higher gear.

With certain exceptions, a rear axle ratio requiring fewer engine revolutions per mile will give better economy. Car makers often list a standard axle ratio in the spec sheet, plus an economy ratio and a heavy-duty ratio intended for towing. The economy ratio will be the lowest number and the heavy-duty ratio will be the highest number.

The exceptions usually arise when you don't use the car the way it is intended. For example, if you buy the economy rear axle and then use the car to tow a travel trailer you will feel that the gearing isn't right. On even slight upgrades you will have to shift down to keep the momentum. Even on the level you will drive constantly with the accelerator pedal smashed to the floor and you will get poor economy. A higher numerical axle ratio lets the engine turn more revolutions per mile and is equivalent to shifting down a little bit. The engine can now operate comfortably at a middle RPM and you don't have to floorboard the throttle all the time. Gas mileage will improve.

Another effect of rear axle ratio is in climbing grades. An example earlier showed that maintaining cruising speed uphill gave the best economy. However, that car was set up for performance with a rather large engine and an axle ratio of 3.55. It was no sweat for that car to hold cruising speed up a typical highway grade.

With a lower numerical axle ratio such as 2.56 or a smaller and less powerful engine, the situation is different. To hold road speed on an upgrade may require shoving your foot in the carburetor. There went your economy. In this case you are better off letting the speed gradually drop as you go uphill.

If your car is a slug on performance and will barely stay ahead of its own shadow, then let it slow down on the uphill grades.

Other than that, the rules apply generally to any kind of car.

THERE'S JUST ONE THING I DON'T UNDERSTAND

Gee, that's great if there's only one thing you don't understand. What is it?

What do I do?

If you are not planning to enter any economy runs and are not otherwise serious right down to the last drop of fuel, you can sort of play it by ear. Just by following the general rules given on page 15 you can make a worthwhile saving in operating expense.

It does seem like a good idea to measure the gas mileage you are getting as you now drive, to get a baseline. Then put some of the basic principles to work, practice a while, check your mileage again. You'll probably find out that this book didn't cost you anything and your next tune-up is also free.

If you are the type who accepts economy as a challenge or if, sadly, it is an economic necessity, then go at it a little more seriously and scientifically. Try all the tricks, explore different routes and times of departure on your way to work. Install a vacuum gage on your dash and learn to heed what it tells you.

Those suggestions are just in reference to the driving methods discussed in this chapter. We have a lot more to talk about: Things to buy and not buy, modifications you can do or have done, tire selection, buying for economy, selling for the most money, and saving bucks on long trips.

The weight and wind resistance of this camper requires a whole set of extras on the truck, just to drive safely around the block. The only highway economy from a rig like this is saving motel bills.

Towing a lightweight trailer such as this, which doesn't stick up above the El Camino tow car will not have a major effect on highway gas mileage or car performance. To use this trailer on the annual two-week vacation, the car doesn't need a special axle ratio or other equipment. Just conservative driving.

Nine ways to get better fuel mileage

Would You Believe 376.59 Miles Per Gallon?—The following paper is reproduced here through courtesy of the Society of Automotive Engineers, Inc., and the author Mr. D. L. Berry of Shell Oil Company. It was delivered to the West Coast Meeting of the SAE in 1974. The activities of the Shell Mileage Marathon(TM) are recreational but there is a serious note behind it all and some very good information about what makes good mileage.

THE OBJECTIVE of this paper is to discuss techniques of motor vehicle operation which can contribute to reduced fuel consumption. Although some drivers may continue to be concerned only with the problem of finding gasoline to keep their tanks full, increasing prices and the probability of long-term shortages should provide the incentive for others to take serious interest in the fuel economy of their cars and in methods of fuel conservation.

Certainly, there is no lack of discussion and advice on the subject of saving fuel, and special "Tips to Motorists" are available from many sources. Rather than simply list a number of mileage improvement suggestion, a review of some of the fuel saving methods used over the years by the mileage experts competing in the Shell Mileage Marathon(TM) is presented.

MILEAGE MARATHON

This event, sponsored as a recreational activity by personnel of the Shell Development Co. Research Laboratory at Wood River, Illinois, originated in 1939. Since that time, 26 mileage competitions have been held involving a total of more than 700 entries. An early description stated that the Marathon was "a friendly contest to learn which of the staff who, through his uncanny skill, superior knowledge, or slick cunning, is able to drive his car farthest with the gasoline provided."

Over the years, interest in this contest has been maintained by the strong competitive spirit of the participants and the challenge of striving to reach the ultimate in fuel mileage. It might seem that the ultimate may have been reached in 1973 with 376 miles/gal. But now, the goal is 400 miles/gal.

Fig. 1 is an illustration of the maximum miles per gallon attained in all contests to date. The 100 miles/gal. barrier was exceeded in 1949, and later the 200 and 300 miles/gal. marks also were surpassed. A listing of cars achieving new records is given in Table 1. It will be noted that in addition to miles per gallon data, the car performance results are shown also in terms of ton-miles per gallon (TMPG) in recognition of the important effect of vehicle weight on fuel consumption.

The methods for improving fuel mileage can be classified into the three major areas as indicated below:
1. Decrease rolling resistance and aerodynamic drag.
2. Increase engine efficiency.
3. Use efficient driving technique.

To illustrate the importance of each of these, examples of their application will be described in connection with the Marathon performance of four of the cars. All of these were specially modified and operated in the Unlimited Division of competition. As noted from Table 2, the contest rules for this classification had very few restrictions.

In the later descriptions it will be seen that although some of the modifications and techniques were too specialized to be practical for use by the average car owner, the basic principles can be applied in various ways to most vehicles.

1947 STUDEBAKER

One of the cars of particular interest was a 1947 6-cyl, manual transmission Studebaker (Fig. 2) which attained 149.95 miles/gal. in the 1949 contest (1)*. A special feature of this car was the home-built molded Plexiglas bubble-nose section designed to minimize air resistance. The bumper, grille, and underside of the car also were streamlined by installation of a smooth-surface covering.

As shown in Fig. 3, air resistance, which varies in proportion to the third power of the speed, represents a major part of the road load-power required or total resistance of a vehicle. Rolling resistance (tires, wheel bearings, driveline, etc.) also increases with speed but at a lower rate. To avoid the adverse effects of air resistance, vehicle speeds in most Marathon competition were kept below 25 MPH; however, the test course used in 1949 included hilly terrain, and at one location, the Studebaker could attain speeds up to 55 MPH in downhill coasting.

*Numbers in parentheses designate References at end of paper.

Fig. 2 - 1947 Studebaker—149.95 miles/gal.

This was considered to be sufficient justification for the substantial effort involved in developing special streamlining. It should be noted that closed courses were used in these contests so that even with hilly terrain, there was no net change in elevation from the start to finish point. In recent years, increased traffic density has required a change to a limited-access, level-road test course.

With respect to the air resistance factor, the average driver has little opportunity to utilize streamlining factors other than to take advantage of the decreased drag effect of keeping windows closed and to avoid use of a car top carrier. High speed operation is limited of course by the 55 MPH federal speed law, but fuel economy can be improved further in most current automoblies by operating in the 30-40 MPH speed range.

Over the years of Marathon competition, considerable effort has been applied in exploring and developing methods for the reduction of rolling resistance and the improvement of drive-line efficiency. With reference again to the 1947 Studebaker, the following modifications relate to rolling resistance and chassis friction:

1. Wheels—Original 15 in size replaced by 16 in; precision balance; front wheels aligned for minimum side drag.
2. Tires—Original 5.50 x 15 size replaced by 7.00 x 16; treads buffed smooth, leaving narrow center strip; tire pressure: 110 PSI.
3. Rear axle and transmission—Original hypoid gears replaced by spiral bevel type (gear ratio changed from 4.56:1 to 3.9:1); freewheeling overdrive unit installed.
4. Lubricants and seals—Wheel bearing grease seals removed; light oil used for bearing lubricant; light oil used in transmission, rear axle.
5. Brakes—Adjusted for zero drag when released.

When prepared for contest operation and parked on a level surface, the Studebaker and some of the other highly modified cars could easily be set in motion by the touch of a finger tip. In contrast, a modern 4500 lb. car with soft, wide-tread tires, hypoid-type rear axle gears, and perhaps dragging brakes would be difficult to push even on a slight downgrade.

Maintaining proper tire pressure is an effective way to decrease rolling resistance in any car. In Fig. 4, the effect of tire pressure on fuel mileage is shown for the 1947 Studebaker (upper curve) and also for a current-type 5300 lb. car fitted with H78 x 15 size tires. It is shown that increasing pressure from 15 to 32 PSI (maximum recommended) at 45 MPH improved the fuel mileage by 9.5%. The use of radial-ply tires can result in additional mileage gains. For all cars, it is important to consider that driveline friction and fuel mileage can be affected adversely by factors such as improper wheel alignment or bearing adjustment, a dragging parking brake, and heavy gear lubricants.

Following the 1949 contest, tests were made to determine the separate effects of the various equipment modifications in the Studebaker. Some of the results are shown in Fig. 5 in terms of comparative miles per gallon under level-road load conditions.

1924 CHEVROLET

In 1952, the winning car was a 1924, 4-cyl Chevrolet (Fig. 6) which attained 168.47 miles/gal, a world's record which was not broken for 15 years.

The inherent low rolling and driveline friction of this car, enhanced by the over-

Fig. 1 - Maximum miles per gallon—Marathon competition

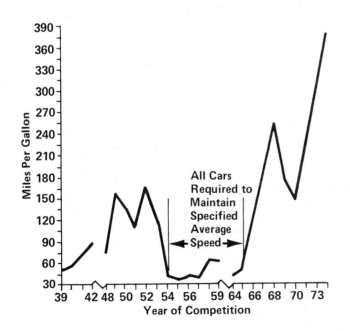

Table 1 - Marathon Mileage Records

Year	Car	TMPG	Miles/Gal.
1939	1933 Plymouth	71.6	49.7
1949	1947 Studebaker	240.0	150.0
1952	1924 Chevrolet	206.4	168.5
1968	1959 Fiat	186.9	244.4
1969	1953 Studebaker	273.6	109.5
1971	1956 Austin-Healey	323.9	156.1
1972	1959 Opel	306.7	297.7
1973	1959 Opel	451.9	376.6

Table 2 - Marathon Rules—Unlimited Division

Engine Modification	Vehicle Modification	Driving Technique
No restrictions other than that the engine must have been available in the car model.	No restriction other than that the car must weigh a minimum of 2000 lb. or exceed manufactured weight.	No restriction or time limit.

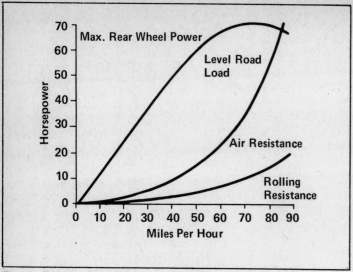

Fig. 3 - Relations between power available and car resistance

Fig. 4 - Effect of tire pressure on miles per gallon

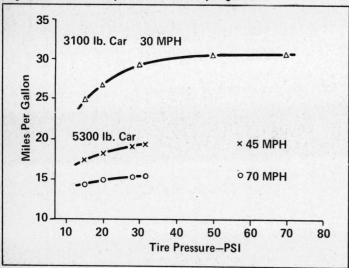

Fig. 5 - Effect of equipment modifications on miles per gallon—1947 Studebaker

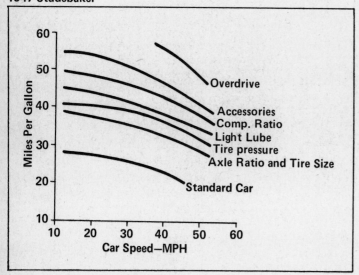

size smooth-tread tires (33 x 5 size), provided for outstanding coasting ability. As a demonstration of the wheel bearing efficiency, the front wheels, when jacked clear of the ground and given a firm spin by hand, would continue turning for 30 min. There were no front wheel brakes. The standard equipment spiral-bevel rear axle gears allowed the rear wheels to coast freely. A freewheeling unit (1933 Chevrolet-type) mounted at the rear of the transmission facilitated coasting operation and clutch disengagement. It might be recalled that freewheeling devices were offered for several car makes in the early 1930's as a fuel saving accessory.

To decrease air resistance, a special body was constructed, including a streamlined nose section and an undercar enclosure. The frontal area was reduced substantially as compared with the original 2-door coupe in which the windshield alone had an area of more than 6 ft.^{2} (Fig. 7).

The engine of the 1924 Chevrolet (Fig. 8) was a 4-cyl, overhead valve-type of 154 in.^{3} displacement. As would be true for many of the special cars used in the contests, careful attention was given to the mechanical condition of all engine components, and various modifications and special adjustments were made to assure optimum performance. Factors considered in the Chevrolet are:

1. Engine block—New sleeves and pistons installed with 0.003-0.004 in. running clearance.
2. Pistons—Light weight aluminum-alloy, two rings per cylinder.
3. Cylinder head—New valves, springs, special rocker arm needle bearings.
4. Compression ratio—Increased from 5:1 to 6:1 (limited by crankshaft durability).
5. Crankshaft, con. rod bearings—Checked for alignment, bearings hand-scraped for precision contact and clearance.
6. Induction system—Special intake manifold: Water-Jacketed, insulated; special carburetor: Small venturi, adjustable metering set for optimum lean A/F; exhaust-heated air duct (standard for 1924).
7. Ignition system—Modified distributor: Spark advance tailored for best fuel economy during acceleration; spark

plugs: special, 0.060 in gap; ignition coil: Converted to 12 V system.

8. Temperature control—Radiator completely covered, fan removed; engine insulated to reduce heat loss.

9. Engine lubricant—SAE 5W grade.

Generally, in the preparation of Marathon cars, an important consideration was the elimination of power-absorbing accessories. In the case of the Chevrolet, which had only basic accessories, the only change was removal of the fan blades. The water pump and generator (set for low charge rate) continued to operate. Because of the very low fuel flow and heat input in Marathon operation, engine cooling was not a problem, and the fan was not necessary.

In typical current automobiles, the possibilities for decreasing accessory power losses would include the following:

1. Minimize use of air conditioner and avoid control settings that permit continous compressor operation.

2. Install flexible-blade or viscous-drive fan.

3. Minimize electrical loads by avoiding excessive cranking (causes increased alternator load to restore battery charge), high blower speeds, and unnecessary lights.

4. Check drive belts for proper adjustment.

Fig. 6 - 1924 Chevrolet—168.47 Miles/gal.

Fig. 7 - 1924 Chevrolet—before modification

Fig. 8 - 1924 Chevrolet—engine compartment

MARATHON DRIVING TECHNIQUE

In an earlier illustration, it was observed that the Studebaker could achieve about 55 miles/gal. under steady cruise conditions on a level road. Similarly, the fuel economy for the 1924 Chevrolet was 50 miles/gal. under the same conditions. The large increase from this level to the 150-168 miles/gal. range attained by these cars under the conditions of contest performance can be explained by the driving technique employed.

Using typical level road data for the Chevrolet, Table 3 shows that for steady cruise conditions, operation at 50 miles/gal. is equivalent to a fuel flow of 75.7 ml/mile. In contrast, with an operating sequence using full throttle, high gear accelerations from 3 to about 15 MPH, then coasting with engine off to the initial speed of 3 MPH, the fuel consumption per mile was reduced to only 23 ml (equivalent to 164.5 miles/gal). Typical test runs indicated that for level-road operation six acceleration/coast cycles (0.16 miles each) were required for each mile traveled.

The remarkable gain in fuel economy obtained by this procedure is based on the well-known relationship (Fig. 9) showing that for a given speed, an automobile engine operates at best fuel economy (minimum BSFC) at near full throttle (full load). With this method of driving, the engine is operated intermittently and only under the most efficient conditions. The energy supplied to the vehicle is utilized in power-off coasting at zero fuel flow. It should be noted that because it was essential for the engine to operate at very lean A/F throughout the selected acceleration range, the normal practice was to deactivate the acceleration pump and power enrichment systems of the carburetor.

This technique, which could be called "momentum-driving" was first used in the 1940 contest and has been applied extensively in all later competitions. This procedure is most effective in cars with manual transmissions which permit acceleration from low speeds in high gear with no "slippage."

Obviously, this specialized procedure would not be at all practical for normal use; however, it is of interest inasmuch as the very high mileages attained in the Marathon competition would not have been possible by any other more conventional driving method.

THE 300 MILES/GAL. CARS

In the Marathon activities to date, two cars have reached the 300 miles/gal. level. One of these is a 1959 Fiat 600 (Fig. 10). This subcompact size car, with a 39 in.^{3}, 4-cyl rear-mounted engine, set a new record of 244 miles/gal. in 1968 (2) and took second place in 1973 with 304 miles-gal. The features of low frontal area and light weight contributed substantially to the good mileage performance of this vehicle.

As one means of reducing driveline friction, a general practice in highly modified cars has been the installation of specially-built freewheeling clutches in the drive wheel hubs. In the case of the Fiat, the clutches were incorporated in the transaxle driveshafts (Fig.11). These devices permitted normal forward drive

Fig. 10 - 1959 Fiat model 600—304.17 miles/gal.

Table 3 - Effect of Driving Technique—1924 Chevrolet

	Steady Cruise	Accel./Coast
Speed		
MPH	50	3-15-3
RPM	1900	250-600-0
Fuel/mile		
gal.	0.020	0.006
ml.	75.7	23
Miles/gal.	50	164.5
Fuel/accel., gal.	—	0.001
Miles: accel. + coast	—	(6 accel./mile)

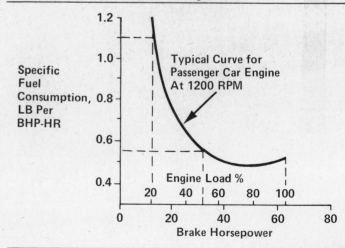

Fig. 9 - Relation between BSFC and engine load

Fig. 11 - Transaxle freewheeling unit

The symbol *ml* used in this portion of the text is an abbreviation for *milliliter*—one thousandth of a liter. If you remember from school days that a thousandth of a liter is one *cc* or *cubic centimeter,* you are still O.K. Both *ml* and *cc* are measures of volume and they are practically identical. The difference is of no consequence to average citizens driving around in their cars.

Fig. 12 - Freewheeling unit—disassembled

Fig. 13 - 1959 Opel—376.59 miles/gal.

Fig. 14 - 1959 Opel—rear axle and chain drive

but allowed the rear wheels to turn freely during coasting, and thus no energy was lost in turning the transmission shafts or main drive clutch. A disassembled unit is shown in Fig. 12. Level-road coasting tests showed that the effect of these units was to increase the coasting distance by 21%.

As stated earlier, the present Marathon record is 376 miles/gal. set in the 1973 contest in a greatly modified 1959 Opel (Fig. 13). Perhaps the performance of this car will be more believable after observing its propulsion system. As shown in Fig. 14, the rear wheels are mounted near the center of a special rear axle shaft supported by roller bearings. Power from the transverse-mounted engine was transmitted by a chain drive directly from a crankshaft sprocket to the axle shaft. A clutch at the axle sprocket permits disengagement for coasting.

Smooth-tread aircraft tires used on all wheels were inflated to 200 PSI. It is of interest to consider that in this car, the fuel consumption per mile was slightly more than 10 ml or volume of about two teaspoons.

This brief review of the performance of four Marathon cars has emphasized the degree of effort required to attain mileages at the 100 plus levels. Obviously, the extent of vehicle modifications and the operating techniques used in these "all-out" record attempts are not practical for the normal range of conditions which are of concern to most drivers.

For today's average driver, vehicle operation is influenced largely by factors such as traffic flow, congestion, speed limits, stop signs, terrain, and weather. The following section discusses fuel saving recommendations based on consideration of the characteristics of a typical modern automobile (unmodified) and the operational limitations just mentioned.

NINE WAYS TO SAVE FUEL

The title of this paper suggests that there are nine ways to save fuel. Certainly there are many more, and the list presented in Table 4 could be expanded considerably. These items, all of which are under the driver's control, require no vehicle modification, and if carefully applied, can be very effective. As has always been true for Marathon operation, driving to conserve fuel should not be a casual matter; it requires constant attention and the evaluation of every action in terms of its effect on fuel consumption.

As an illustration of the miles per gallon attainable in unmodified passenger cars when careful driving habits are applied, Table 5 is a tabulation of Marathon data showing the best miles per gallon performance of cars in the Sportsman Division of competition for 1971, 1972, and 1973. As noted from the

---

Table 4 - Nine Ways to Save Fuel

1. Use moderate speeds (best cruise speed is 30-40 MPH).
2. Drive at smooth steady pace (when traffic permits).
3. Accelerate slowly, allow automatic to upshift.
4. Anticipate stops, minimize braking.
5. Prolonged idle for warmup not necessary.
6. Limit extensive idling, stop engine.
7. Be sure parking brake is released fully.
8. Minimize electrical loads and use of air conditioner.
9. Consolidate short trips, plan routes in advance, avoid heavy traffic.

Table 5 - Best Mile per Gallon Performance—Marathon Sportsman Division*
Unmodified Passenger Cars

Weight Class, lb.	Contest Year	Car Year and Make	Running Weight	Transmission Type	Miles/Gal.
2400	1973	1969 Volkswagen	2285	4-speed	49.4
	1972	1972 Datsun	2400	4-speed	44.1
	1971	1969 Volkswagen	2400	4-speed	49.9
3000	1973	1972 Pinto	2725	4-speed	35.5
	1972	1969 Volkswagen	2742	4-speed	44.2
	1971	1968 Peugeot	2995	4-speed	38.9
3500	1973	1970 Volvo	3175	4-speed	31.9
	1972	1968 Mustang	3600	Automatic	30.9
	1971	1968 Volvo	3390	4-speed	36.8
4000	1973	1973 Dart	3855	Automatic	24.7
	1972	1969 Barracuda	3984	Automatic	24.2
	1971	1968 Chevelle	3990	Automatic	25.8
4500	1973	1970 Dodge	4470	Automatic	24.6
	1972	1969 Plymouth	4583	Automatic	23.5
	1971	1969 Ford	4655	Automatic	24.4
5000	1973	1965 Pontiac	5005	Automatic	22.0
	1972	1965 Pontiac	5140	Automatic	21.4
	1971	1966 Buick	5320	Automatic	18.7

*See Appendix B for a description of contest rules for this division.

summary of the rules for this classification (appendix B), no vehicle modifications were permitted, and the cars had to maintain an average speed (without coasting) of 30 MPH.

In Table 6, mileage test data for a typical late model automobile, measured under a variety of operating conditions, illustrate the relative effects of speed, acceleration, engine idling, and air conditioner operation. It should be recognized that tests in other vehicle types would show different results. For example, other data sources indicate that operation of an air conditioner might cause a 2 miles-gal. loss, or under hot temperature conditions in heavy traffic, as much as a 25% loss. Fuel consumption at idle in a large V-8 engine can be as high as 0.8-1.0 gal./h.

The most meticulous fuel saving efforts by a driver can be completely nullified by engine malfunctions, improper adjustments, or neglect of simple maintenance and inspection items which should receive careful attention in every vehicle. It will be recognized that these factors relate to the general categories of "Improved Engine Efficiency" and "Decreased Rolling Resistance."

Proper functioning of components associated with emission control systems is especially critical in relation to fuel economy. It is particularly important that temperature switches, solenoid valves, and orifice valves used to control the application of distributor vacuum spark advance be checked for proper installation and operation. Because of extensive use of engine vacuum in the operation of these units, an undetected vacuum leak can be detrimental to performance and economy.

SUMMARY

The attainment of fuel economy at the 150 miles/gal. level is considered to be a fairly easy accomplishment by some of the experts who compete in the Shell Mileage Marathon. A brief review of the performance of four record mileage cars shows that the fuel saving methods which have been developed over a period of more than 25 years involve various vehicle modifications and detailed attention to many factors all of which contribute to the basic objectives of decreasing rolling resistance and air drag, improving engine and power train efficiency, and using the most efficient driving technique. For the highly modified cars, it is evident that the driver must accept some compromises with respect to vehicle comfort and convenience, ease of operation, and speed range capability.

With respect to the application of fuel saving procedures in the case of an average driver with a typical modern automobile, the possibilities for mileage improvement are, of course, limited by vehicle factors such as size and weight, engine and transmission type, power accessories, and emission control systems. Practical considerations rule out any major vehicle modifications and in many late model cars, the range of engine adjustments is restricted as required to maintain normal emission standards.

Despite these considerations, and the conditions of normal service which involve variables of traffic, weather, temperature, and terrain, it is believed that any car can be operated with improved efficiency, and there is no doubt that the combination of serious and consistent application of good maintenance practices and careful driving habits can result in significant fuel savings.

Table 6 - Effect of Operating Conditions on Miles per Gallon

Operating Condition	Miles/Gal. Without A/C	Miles/Gal. With A/C
25 MPH cruise	18.4	17.0
50 MPH cruise	17.2	16.3
70 MPH cruise	14.4	13.6
0-50 pt acceleration	13.4	13.0
0-10 wot acceleration	12.0	11.3
50-65 mwd acceleration	11.6	10.5
0-30 10 stops/mile	5.7	5.5
60 s idle + 5 mile at 50	16.2	—
60 s idle + 1 mile at 50	13.3	—

Table 7 - Maintenance and Inspection Items

Engine	Emission Devices	Chassis
Dwell, spark timing	Intake air heat	Tire pressure
Idle speed (incl. fast idle)	Spark control units valves, orifices, switches	Brake drag (park brake/adjusters)
Auto. choke: adj., operation		Lubricants, fluids
Spark plugs, wires		
Engine temp.	EGR valve	
Drive belts	Vacuum hoses, leaks	
Air filter	PCV system	

EDITOR'S NOTE

Abbreviations used in Table 6 are commonly used and understood in the automobile and related industries but may require explanation for ordinary citizens. The expression *pt* acceleration means *part throttle* acceleration. Accelerating at *wot* means *wide-open throttle*. Testing at *mwd* which means *maximum without downshift* is done on automatic transmission cars where you accelerate as hard as possible in a particular gear but not hard enough to cause the transmission to downshift into the next lower gear.

APPENDIX A

Table A-1 - Summary of Marathon Mileage Data—Four Cars

Car Year, Make	Engine Data	Marathon Year	Contest Weight, lb.	TMPG	Miles/Gal.
1947 Studebaker	6-cyl, 170 in.^{3}	1949	3206	240.0	149.95
		1950	3345	224.5	134.18
		1953	3400	189.2	111.30
1924 Chevrolet	4-cyl, 154 in.^{3}	1949	2509	198.9	158.36
		1950	2500	148.1	118.51
		1951	2450	138.0	113.00
		1952	2450	206.4	168.47
1959 Fiat	4-cyl, 39 in.^{3}	1967	1600	138.5	173.16
		1968	1530	186.93	244.35
		1972	1470	184.42	250.91
		1973	1640	249.42	304.17
1959 Opel	4-cyl, 91 in.^{3}	1970	2020	142.28	140.87
		1971	2210	245.50	222.17
		1972	2060	306.66	297.73
		1973	2400	451.9	376.59

APPENDIX B

Table B-1 - Marathon Rules

Engine Modification	Vehicle Modification	Driving Technique
Modified Division—Divided into classes according to vehicle weight and transmission type		
Modifications limited to carburetion and ignition systems	No modifications allowed Maximum weight restriction Maximum 40 PSI tire pressure	No restriction, but must maintain 30 MPH average
Sportsman Division—Divided into classes according to vehicle weight		
No modifications Must be tuned to manufacturer's specifications	No modifications allowed Maximum weight restriction Maximum 40 PSI tire pressure	Car must be in gear at all times (no coasting) 30 MPH average

EDITOR'S NOTE:

Sharp-eyed and technically-minded readers may notice an apparent conflict between this SAE paper and the statement earlier in this book that wind resistance doubles when vehicle speed doubles. This SAE paper says "air resistance . . . varies in proportion to the third power of speed . . ."

If you are not interested in haggling about it, just remember that wind resistance goes up a lot when you drive faster, and don't bother with the rest of this note.

Both statements are correct but obviously are from different standpoints. Wind resistance measured in pounds of force against the car increases with the square of speed. When allocating vehicle horsepower to the various forces it must overcome, as done in the SAE paper, the term *air resistance* is taken to mean the amount of vehicle power needed to overcome air resistance. On that basis, horsepower needed to overcome air resistance increases with the third power of speed.

REFERENCES

1. R. J. Greenshields, "150 Miles Per Gallon is Possible." SAE Journal, Vol. 58 (March 1950), pp. 34-38.

2. D. C. Carlson and H. D. Millay, "Mileage Marathon from 50 to 244 MPG." Paper 700532 presented at SAE St. Louis Section Meeting, December 1969.

3 ECONOMY ON YOUR VACATION TRIP

One of the Great American Customs is the annual 2-week motorized vacation. The whole family piles into the car or wagon, Mom and Pop close their eyes to what the cost of the trip is going to do to the family bank account, and away they go. Champions at this sport manage to drive through 30 or 40 states and "see" 8 or 10 tourist attractions all in one glorious 2-week traveling binge.

The energy shortage so far hasn't changed this very much. People now tend to visit fewer places and stay longer at each place, but the vacation trip is still on.

If you are going to invest your time and money in a long trip, it makes sense to get the most return for your investment. Naturally you will drive for best fuel economy but there are other considerations. Time spent walking the sidewalks of a strange town while the local auto repair shop is working on your car—or waiting for parts—is not a fun vacation. The bill for unexpected repairs has put an end to many trips and the kids never got to see Disneyland.

TRIP CHECKING

Thirty minutes spent checking a car out from stem to stern prior to taking a trip can often save a lot of time and a lot of money. Even without a long trip, the items listed here should be checked at least every six months so you'll be able to nab trouble before it nabs you.

When making a routine check under the car or under the hood, the attitude to take is that something is wrong—it's your job to find it. Chances are, if you assume nothing is wrong you won't find anything.

Planning, saving, spending, and dreaming for a year or more are all preliminary to the great vacation. If you are going to do it, do it right and bring back happy memories.

The trip began with a bad radiator hose and ended like this. Check your car before the trip.

Before your trip—and occasionally even for driving around town—check the fan and accessory drive belts for slack. New belts tend to stretch a bit during the first few weeks and should be re-checked.

Inspect the under side of the belts—that's where failure begins.

BELTS

Air conditioning, water pump, power steering and smog pumps are all driven by rubber and fabric belts. These must be checked for tightness and for cracks that could lead to failure. Check the belt for tightness by hooking a finger around the belt about mid-way between two pulleys and pulling outward. If a belt is too loose it will move outward an inch or so; a belt too tight won't move at all. The belt should be loose enough to be pulled outward about 1/2 inch with a modest tug. A factory service manual for your car will give the exact amount of deflection on all drive belts. A too-tight belt can cause a bearing failure in the unit it is driving such as an alternator or power-steering pump. A too-loose belt will slip and could cause overheating or very poor operation of the driven unit. A loose belt will also fail in a shorter period of time than one adjusted to the proper tension.

Inspect the belts for any nicks, cuts or cracks—sometimes called checking. Any irregularities can lead to a belt coming apart at that point. Don't just check the outside of the belt, check the inside surface that rides in the pulley. Bump the engine over with the starter so you can check all the belt—not just the part easily seen while you are peering down on the engine.

When installing a new belt there are a couple of little tricks you should be aware of. Never, ever force a belt over a pulley with a screw driver. This can tear some of the cords within the structure of the belt and the belt will fail prematurely. Loosen the mount of the alternator or other driven device until the belt will slip on easily. Also keep in mind that a new belt will stretch so you can snug it right up to the limit of tension. Within several hundred miles the belt will loosen up and can then be retightened. On the other hand an old belt will not stretch. It tends to be hard and brittle and if you must reinstall one do not tighten it as tightly as you would a new belt.

When replacing a belt, make sure you replace it with one of the correct size—not only length but also the correct width so it will fit down in the pulley correctly. In many cases belts are designed specifically for an engine. It may have a steel cord within the belt instead of a fabric cord. The belt may be wrapped with a fabric

Never install a new fan belt by levering it over the side of the pulley. This can rupture the cording in the belt and cause early failure. Look around and you will see an idler pulley which can be moved to create enough slack to remove and install a belt—or else the driven accessory can be moved by loosening its mounting bolts. The old fan belt can be saved for a spare if it is not totally destroyed.

HOSES

A blown radiator or heater hose on the highway on a hot day can burn an engine up before you can turn the key off and coast to a stop! Sometimes you can drive a long way with a bad and leaking water pump, but when a hose lets go and the coolant leaves the engine in a hurry, pistons seize, bearings lock up and oil in the crankcase can actually boil. When this occurs you are looking at a towing bill, a new or rebuilt engine and a long period of time without a car. Don't gamble the cost of a new hose against the cost of an engine.

When the engine is cold, check both the upper and lower radiator hose and both the supply and return hoses for the heater. All rubber hoses should feel firm, alive and flexible—not flimsy like a rag or brittle. When a hose becomes hard, a vibration or someone innocently bumping against it while checking oil can split the hose. Check the underside of the hose for chafing against any part of the car. If a hose is routed near an exhaust manifold, the added heat can shorten the life of a hose. Because heater hoses are longer and smaller in diameter than a radiator hose they tend to move more and they often fail before the radiator hoses. Check the heater hoses where they enter the firewall and where they connect to the engine. Don't be bashful. Move the hose, squeeze it. It is far better that it fail while cold and in your driveway than to have it let go while you are cruising along a lonely highway.

On many late V-8 engines the upper

and designed so there is some slippage during quick accelerations to prevent the belt from turning over and running upside down in the pulley at high speed. This is especially true of many modern V-8's. When possible, all drive belts on your car should be replaced with belts of the same brand and size as those which came on the car when new. Here is a case where the discount house should be bypassed and an item purchased for a higher price. If the belt lasts longer and is trouble-free that is far better economy than buying a lower-cost belt which has a short life or whips itself off the pulley. Those tow bills from remote areas can be expensive, so save bucks by spending cents—get the proper belt for the application.

Change the belts before they break. Carry the old ones in the car with your spare tire. That way you may be able to continue your trip if a belt fails . . . provided you have tools along to change the belt.

A replacement belt should fit properly in the pulley or sheave. If it bulges out like this, it's too big for the groove and the belt will wear out quickly.

A belt that's too narrow drops down into the groove like this and tends to fail due to overheating.

Yup. This is the way it should look. This belt fits properly.

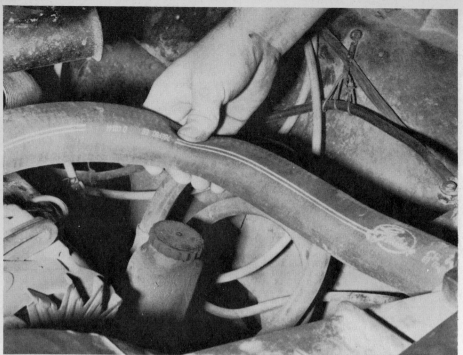
If the radiator hose feels soft and compliant, it's probably in good shape.

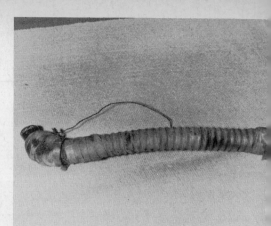
Wrapping a hose with tape and holding it in place with wire may be a good emergency fix on the road—but don't start out that way.

Check the brake pedal. If it goes most of the way to the floor, check the brake fluid and have the brakes adjusted.

Check brake fluid by lifting the top of the reservoir. Use only the brand and type of brake fluid specified in your owner's manual.

hose is very long and may have a piece of metal tubing joining two lengths of hose to make the connection between engine and radiator. This can cause trouble of another kind. All that weight jiggling around mile after mile can cause the inlet neck on the radiator to crack and pull away from the upper radiator tank. When checking the upper radiator hose always check the upper tank inlets for coolant weeping or seepage.

If you plan to keep a car for a very long time—or just want the very best heater and radiator hoses that money can buy, check the parts counter of a large truck dealer. Diesels use a fairly new expensive silicone-rubber hose. It will withstand great pressures and temperatures and give extremely long life. Compared to a regular hose, the silicone-rubber item is a far better buy if you plan on driving a lot or keeping the vehicle a very long time—which is why they are used by truckers. The hoses come in a great variety of lengths and diameters—although you might not be able to find one to fit applications that require complicated molded hoses.

FLUID LEVELS

All fluid levels in a vehicle should be checked before leaving on any extended trip. These include brake fluid in the master cylinder, engine oil, battery and radiator water, transmission and rear-axle lube, and the power steering. There are a couple of things you should know before trying to make a 30-minute job into a 10-minute job. First and foremost the fluid levels should be checked when the vehicle is level. Failure to do this can lead to some real problems. For instance if you determine the axle lube is low and add some while the vehicle is not level, the excessive fluid could seep out past the axle seals and run into the rear brakes after the fluid expands from heat. Unless otherwise specified check fluid levels when cold. This keeps you from winding up with a painful burn or unnecessary discomfort from heated components.

In the case of the automatic transmission, the engine should be warmed up and the transmission placed in PARK and the fluid level checked with the engine running. If the automatic transmission is low on fluid, add it very slowly and

Check engine oil with dipstick on side of engine. Pull it out, wipe it clean, put it back in, pull it out again and check the level against marks on the stick.

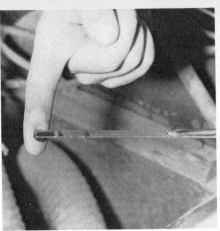

Usually dipsticks have at least a "low" mark and a "normal" mark. Usually one quart of oil will bring the level from low to normal.

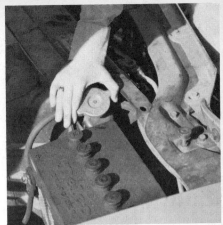

Remove radiator cap and check coolant level. If the engine is hot, turn the cap about a half-turn or until pressure starts to escape and then wait for things to settle down before you remove the cap all the way. Otherwise you may get a face full of steam or boiling water. Then remove battery filler caps and check water level visually, it should come up to the bottom of the filler openings on most batteries.

Automatic transmission fluid level is checked with dipstick near rear part of engine. A funnel with a long spout is used to put in fluid. These are available at auto supply stores.

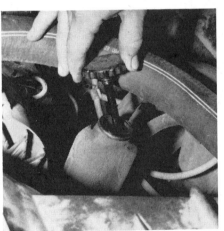

Checking the fluid level in the power steering pump is so simple it should be done each time the engine oil is checked. Use automatic transmission fluid to refill.

then recheck. Never, ever, dump in a full quart of trans fluid and then check the level. What appears to be a very low fluid level on the dipstick can actually mean the transmission needs only a cupful of fluid. If the trans is overfilled it will force out the fluid as you drive. This is super messy because it will coat the underside of the car and land on the hot exhaust system where it will go up in smoke. Automatics have a habit of dumping out a lot of fluid when they are overfilled.

Automatic transmission fluid is also used in the power steering pumps so while you have a can of fluid in hand, check that item. Power steering reservoir filler caps are usually coated with oil and dirt. Be careful not to get dirt into the reservoir when taking the cap off or putting it back on.

If you have a manual transmission, it will have a separate oil supply which is checked and refilled from under the car. The rear axle is also serviced from underneath. Refilling these enclosures is not convenient to do at home and you are probably better off having these checked at a service station when you fill up with gas.

Watch your battery like a hawk in the summer time. Don't overfill but keep the plates covered. Invest in a gallon of distilled water and use only this in the battery. Mineral deposits in ordinary tap water will shorten the life of a battery. Another way of getting what you paid for. Before a long trip, drop by a service station and get them to check the specific gravity of the acid and water with a hydrometer. All of the cells should have equal hydrometer readings. Any low cell is about to "go away" and the battery could fail when you need it most.

UNDERHOOD WIRING

While you are checking fluid levels and the engine is cold and easy to work around, give all of the wiring under the hood a close look. Start with the battery terminals—follow the ground strap to the frame or engine and do the same with the hot lead to the starter. Are the terminals tight and free of corrosion? Is the hot lead free of all obstructions or is it chafing on something? Modern wiring is pretty good stuff, but thousands of miles of driving can cause the wiring to chafe against an inner fender panel or vibration can cause a screw to back out of a terminal. A tiedown can come loose and allow a harness to come close to an exhaust manifold.

I know an incident where a tiedown broke and the wiring harness leading from the alternator to the firewall fell down and became entangled with the steering shaft coupler. On a hard turn the coupler yanked the harness out of the alternator and pulled the oil-pressure sending-unit wire out of the block. The steering wheel was in quite a bind until the wiring finally pulled loose. This could have caused a serious mishap—all because a wiring tiedown failed. A close under-hood inspection would have caught the problem.

LIGHTS

Even if you do not plan to drive at night, always check all lights front and rear on your car. You could be forced into a night driving situation and a patrolman could discover you don't have a brake light in operating order—or a tail light. Replacing a bulb or fuse to fix an item like this is far less expensive and less hassle than any ticket you might get. If you operate the car in the north or eastern areas of the country, do more

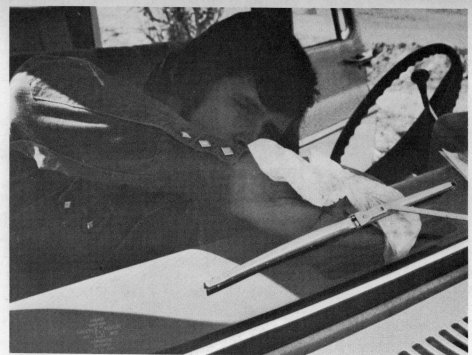

When you find yourself taking a dim view of the situation, clean the *inside* of your windshield. The amount of grease and dust that accumulates there is amazing.

Even in a dry climate you should keep the windshield washer bottle filled. A clean windshield helps tremendously when driving at night or into the sun.

than just check to see that the lights are operating. Make sure rust has not eaten into the receptacle and that everything is firm and in place. The light might work fine while you are standing there looking at it—but vibration from driving could cause the light to go on and off or just go out when you pull out of the driveway. Heavy rust causes many a light to lose its good bulb or ground wire contact.

On most modern cars part of the tail light assembly sticks into the trunk. Make sure you don't dislodge a wire or break off the base of a bulb by throwing a suitcase into a trunk without thought.

TIRES, FRONT SUSPENSION AND EXHAUST—THE DIRTY STUFF

We've lumped these items together because to check any or all of them you had best plan on getting dirty. Dress for the occasion and set aside enough time so you don't have to rush the job.

Tires—Unless you have access to a grease rack so you can get under the car easily, the simple way to check tires is to jack up one wheel at a time. On the front, turn the wheel hard left or right so you can easily examine the tread area for cuts or uneven wear. The rears are easier to see. Rotate the tire slowly. Then with a jack stand under the frame slide under

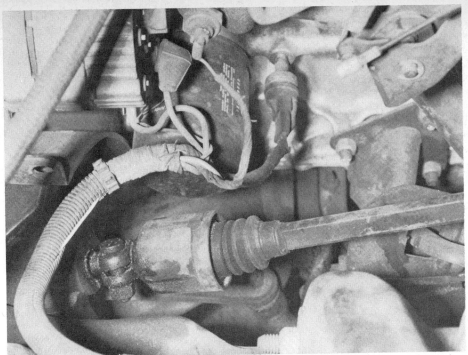

Check underhood wiring to be sure wires aren't rubbing on something and are properly secured. In a car like this one, that bundle of wires dropped down and bound up the steering mechanism.

If any lights don't work after you have *checked them all*, replace burned-out bulbs. If that doesn't fix the problem, check wiring, get professional help.

Each year, thousands of motorists discover that the spare tire doesn't have any air in it. The occasion of this discovery is when some other tire also doesn't have any air in it. Check tire pressures on *five* wheels.

the vehicle far enough so you can see all of the inside sidewall. Examine the sidewall carefully for cuts or blisters. This is a common area of failure because the sidewall simply doesn't get checked often enough. The tire tread can tell you a lot. Uneven wear can tell you the tire has been under-inflated or over-inflated, that the tire is out of balance or that the front suspension is out of alignment. You don't need to be an expert to maintain your vehicle—just know the symptoms of trouble and turn the problem over to a front end specialty shop.

A vacation trip can create tire problems. If you have been running the tires at the minimum recommended pressure to give the car a soft ride and it produces an even wear pattern on the tread; throwing two or three hundred pounds of luggage in the trunk plus taking aboard the entire family will cause the tire to overheat and possibly fail. Even if the tire does not fail, a heavy load and a soft tire can create some unusual and dangerous handling characteristics. The way to eliminate this is to load the car with luggage and all the kids and spend enough time on a familiar road near home to see if the handling has been altered.

If it has, run the pressure up by four

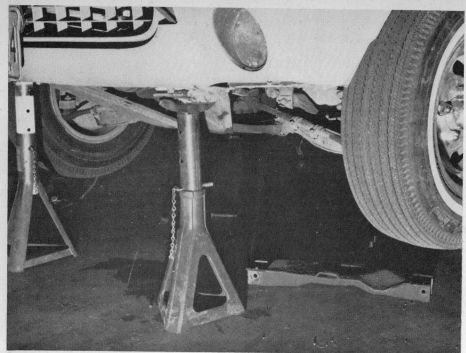

Get the wheels off the ground so you can rotate the tires to inspect them all around and on both sides. Never never get under a car unless you have something sturdy holding it up—such as these jack stands. Don't depend on your trusty bumper jack to keep the car from falling on you.

or five pounds at a local service station. If the car is obviously tail-heavy, bleed two or three pounds of air out of the front tires while adding some to the rear. The handling characteristics of a car can be altered a great deal by tire pressure.

If, after you are fully loaded for a trip, the car handles like a water bed, take some time with tire pressures. Remember the heavy end of the vehicle will require the most pressure. Do not exceed the factory-recommended maximum tire pressure warnings. If tire pressure juggling doesn't make the car handle at the speeds you plan to drive, invest in a set of inflatable shocks for the rear. Get your car handling right before you take your trip.

Front Suspension—All of that linkage under the front of a car which has something to do with the steering of a car is pretty durable stuff, but before a long trip you should give it the once over visually. Grab such things as an idler arm, a tie rod or a steering arm and shove and yank them back and forth. Does there seem to be slop or do you hear a clunk? If you are not sure about how much

This tire has been run with too much air pressure which wore out the center of the tread. The edges of the tread are still in pretty good shape, so the owner wasted perhaps 50% of the cost of this tire. Under-inflated tires wear the opposite way—tread edges go first.

Irregular or unsymmetrical tire wear patterns indicate wheel alignment problems or something requiring attention at a good wheel or tire shop.

movement there should be don't be afraid to seek help at the local garage.

A wife or even a child can move a front tire back and forth enough to let you look for loose tie-rod ends. If the ball joint is loose enough to cause trouble, you will be able to hear or feel the looseness. Make sure there is a cotter key in each tie rod end. A shop manual for your car will give you the wear limits for ball joints. This is a typical area for rip offs, so know how much play is allowable.

While you are under a car checking for anything always look at the brake backing plates or brake calipers for signs of brake fluid leakage. Any kind of seepage means you have a leak and leaks don't go away, they get worse! Trace the leak and fix it or have it fixed. A wheel cylinder could be bad or a fitting could be loose.

While you are rooting around under the front end of the car locate the flexible brake lines that run to the wheel cylinders. Make sure they do not show evidence of chafing against some part of the suspension or don't appear to be nicked or cut. Flexible brake lines are very durable but can rupture in a hurry under pressure if they have been previously weakened in any manner. Steel lines can be damaged by rust, or by a flying object such as a rock. Take your brakes seriously. They are important.

Exhaust—With all components cold check the entire exhaust system. Start with the joint between the exhaust manifold and headpipe. Is there soot or other evidence of exhaust leakage on the inner fender panel or firewall or whatever the nearest component is to the joint? Grab the exhaust pipe and shake it. A little movement is normal, but squeaks and rattles means there is trouble brewing. Using the handle of a screwdriver move down the head pipe to the muffler—rapping the pipe, then the muffler. The pipe should sound solid, so should the muffler. When checking the muffler, rap the bottom of it; this should sound solid. Modern mufflers have a small hole punched in the underside to allow condensation to drain out—other than this hole, the muffler should be solid and free of cuts or rusted out holes. Continue your check of the exhaust system the full length of the tail pipe. Shake it, tap it and look for evidence of leaking.

If you find any weak or rusted-out spots, or any reason to suspect an exhaust gas leak, have it checked by a garage or muffler shop before your trip. Exhaust gas is partly carbon monoxide which can seep into a car in motion and put the occupants to sleep—temporarily or permanently.

THE LEGAL STUFF

Economy takes strange forms sometimes—if you don't have proof of ownership in the vehicle you drive, you could be cited for it. That means inconvenience and possibly a ticket. Regardless of where you drive, you should comply with all of the laws applicable to your area. If in doubt, check the local police office, highway patrol or sheriff. A phone call can fill you in. In most states these days if an officer stops you just to tell you that you have a short in the tail light he will probably ask for a driver's license and proof of ownership. If you and the vehicle don't comply to the letter of the law; you will waste some time and maybe some money.

From time to time you may do something out of the ordinary with a vehicle—haul something very wide or long or maybe tow another vehicle. A simple inexpensive permit can save you the cost of a ticket—plus the hassle and time wasted. Here again, a few minutes on the phone to the local law enforcement office can pay off ten-fold . . . or more.

TROUBLE ON THE ROAD

Sooner or later most drivers have car trouble. The trouble may be as simple as a flat tire or as complex as a seized engine. If and when trouble does occur, an understanding of the problem and confidence in yourself are the two great aids to getting out of trouble quickly and with a minimum of expense.

Don't panic—and if you are on the open highway don't immediately assume the only route is to have the vehicle towed to a local garage or service station! A Johnny-on-the-spot tow truck driver will tell you the car must be towed away or it will be ticketed and then impounded. The argument continues that this will cost far more in the long run since you will then have to pay for the ticket, towing, and an impound or storage fee. This may be true—but it may not be the whole truth. Most states allow a vehicle to be left beside a road for a period of time before being ticketed. Even if that period of time is only three or four hours—a lot of repairing can be done in that length of time. In many cases you can

---

USE NAME-BRAND BRAKE FLUID

A very important quality of brake fluid is its boiling point. Brakes and brake parts get very hot during prolonged use such as going downhill in the mountains. If the brake fluid boils in the lines or brake cylinders, it changes to a vapor rather than a liquid and braking of the car is reduced.

Buy name-brand brake fluid, marked "Heavy Duty," on sale if you can. Don't buy cheap brands whether they are on sale or not. The risk of poor quality brake fluid is not worth the few cents saved.

A muffler should be replaced long before it reaches this condition. Carbon monoxide from a leaking exhaust system can seep into a car and asphyxiate the occupants. Struturally weakened exhaust parts can break loose completely and cause a traffic hazard. The noise can get you a ticket. Fix it up before you travel.

Visually check the area under and around the carburetor. If it is sticky, damp or wet-looking and smells of gasoline, your bomb may be a bomb. If it's a leaky fuel-line connection tighten it. If the leak is from the carburetor, fix it or have it repaired.

Best procedure is never to get your car any closer than this to a tow truck. Pre-trip checking allows you to admire this expensive vehicle from afar.

hitch a ride to the nearest town, buy the parts *and* tools needed for the job—make the repair and be back on the road with less time and money spent than if the car had been towed to a garage. This is especially true if you are in a hurry. Just remember that having a car towed to a garage does not mean the vehicle will immediately be repaired: My experience says you will get a delay *and* a big bill.

Normally, the best plan is to attempt to limp into the nearest town and either make the repairs there or have a mechanic of your choice do the work. *Your choice* is the key here, because a wrecker driver may lead you to a garage that will give him a cut of the profits on the repair bill. This is one reason a repair bill can get pretty high.

Several years ago one of my friends was driving a late-model import up a long grade at night. Without warning a high-pitched continuous squeal came from the engine compartment. The engine temperature began to rise. The immediate thought—while pulling to the side of the road—was that the water-pump bearing had failed and the pump had seized causing the car to overheat. A quick check under the hood revealed the radiator was not boiling over and all belts were in place. The engine was restarted and another check revealed that the "smog pump" belt was making the noise by sliding over the pump pulley which wasn't turning. Obviously the pump was frozen. Even before the ignition could be switched off a tow truck pulled over and backed up to within three feet of the front bumper. Clearly he had a fish. My quick-thinking friend met the tow truck driver before his boots touched the ground.

"I've got a real problem, do you have a pocket knife I could borrow a second?"

"Uh, sure."

With pocket knife in hand the offending belt was cut off. The knife was returned, thanks offered and a tow charge avoided. No, it didn't make the wrecker driver happy.

Take the time to locate the problem if you can before committing yourself to any repairs. In this example, because there was no panic the trouble was located in the dark when the engine was restarted for a few seconds. Once the trouble was located, a temporary repair was made—cutting the belt—because the driver knew the smog pump belt did not drive either the water pump or the alternator. Understanding the problem and confidence was the key.

Let's continue to examine this little incident. Think of the wasted time and money if he'd relied on his original thought that the trouble was the water pump. Think of the consequences of cutting the belt in the dark—and then discovering the same belt was used to drive the water pump. Think about haste making waste!

Be calm. Be logical. Avoiding a towing charge is real economy.

Another guy I know was not quite so clever several years ago when driving across country in an old car. The engine began missing and stumbling and lacked horsepower and response far from town one night. Upon reaching the light of a service station it was discovered that the six-cylinder engine had a bad fuel pump. The pump was obviously leaking and only erratic spurts of fuel were coming from the pump outlet. The pump was changed and the trip resumed. Several miles down the road, all of the previous symptoms reappeared. The car was returned to the service station where the new pump was immediately branded as being faulty. Closer inspection revealed gasoline sucked up by the original pump had sprayed back along the engine and has entered the distributor. The points were ruined. A lot of time was wasted on that incident; it could have been a lot of money. Give your problem some thought.

Freak items can cost a lot of money— either unintentionally or by design. Take the case of the air-conditioning compressor emitting strange sounds. It had to be the compressor. The noise occurred only when the compressor was running and the noise came from that area. The choice was to not run the AC (in Arizona!) or to have the compressor fixed or replaced. Fortunately for the owner of the car he discovered the noise was not the compressor but the fan shroud slapping against the radiator frame. Two screws were missing. When the compressor was switched on the engine began to idle slower and a vibration moved the shroud and made the strange noise. Think about the difference in cost between two sheet-metal screws and replacing an air conditioning compressor! Not enough care in the first analysis could have cost a lot of money.

Knowing how and where to get parts for your car away from home can make a great difference in what repairs might cost and the time it can take to get parts. If you plan a long trip to an unfamiliar area with a rare or exotic car, it's a good idea to do a little homework before leaving and figure where parts might be obtained if you need them.

Keep in mind that in many cases it will be less expensive and far less time consuming if you call a friend three states away and ask him to buy the parts and ship them to you by bus or by air freight rather than call a parts house 100 miles away and count on them sending the right part and getting it to you in less than a month.

Two things you should keep in mind while stopped at the side of the road. Both are elementary, but both can get a problem solved in a hurry. Most all motor homes and campers carry a quantity of water. This comes in handy if you need some of the wet stuff to refill a radiator after making makeshift repairs on a hose. Second, many farmers and ranchers have either a gas or electric welding sets— another case of being able to solve a problem with as little hassle and money as possible.

CARRY SOME TOOLS

You need not carry a 400-pound tool box around with you just to drive back and forth to the grocery store, but it's a good idea to have some tools with you at all times. One medium size screw driver, one small and one medium size adjustable wrench and a pair of Vise-Grips wrapped in a couple of good size rags or some old towels will slip under a seat or fit in a trunk and never be noticed. A full roll of electrical tape will make a lot of repairs until you can get home and do the job right. As an added embellishment to your emergency kit, tie it all up with several feet of 10 or 12 gage automotive wire. A roll of duct tape is also handy.

You can't fix a broken spindle with this little kit but you can tighten some nuts and bolts and wire or tape something back together. Obviously, you can get as elaborate with a tool/emergency repair kit as you want. Spend some time wandering through the automotive section of a local discount house; they might have a sale on tubes of hand cleaner or some other useful item that in the long run could make repairs easier or less messy.

SUPPORT YOUR LOCAL WRECKING YARD

Keep in mind that a parts store or a car dealership are not the only places you can get parts for your car. A wrecking yard at the edge of town might have just the water pump, alternator that you need to get back on the road. Sure, you'll be buying a used item—always an unknown quantity there—but you could be hundreds of miles down the road with the used item before the dealership or parts store opened up or came up with the part you needed by having to order it. Wrecking yards are often open on Saturdays—when parts counters at a dealership may be closed. Many wrecking yard owners live on or near the premises and are often glad to open up at most any hour to make a sale. You may pay new price or higher for the part if the owner knows you really need it.

THE HIGH COST OF A BREAKDOWN

When you're traveling in areas where there are long distances between cities, a breakdown can be very costly. If you have a wrecker sent out and are towed to the nearest repair center, a breakdown on the road becomes very grim. Fortunately, among the thousands of cars that travel the road today, only a small percentage ever meet this fate.

To those who fall into that small percentage, there are long delays and high bills. Legitimate expenses are costly, but

A simple tool kit like this will allow you to make many roadside repairs. Match your tool kit to your mechanical abilities and to the jobs that you are willing to undertake without help. If you are a complete mechanic, carry a more complete set of tools.

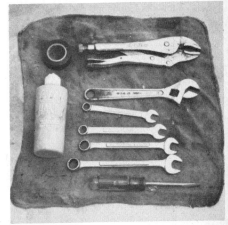

At places like this, it's far better to receive than to give. By application to the manager you can often receive just the part needed to fix your car and continue your trip.

shaky out-of-town shops can quadruple your costs. Not every repair shop along the road is going to take you. Let's just say a good number of them will—through incompetent diagnosis or out and out shifty practices. This emphasizes the importance of careful preparation before you leave. Get the vehicle repaired, look it over yourself and check for the obvious. A dollar item can wind up costing you hundreds of dollars and long delays when away from home. This type of expenditure and delay will definitely upset any vacation or planned trip.

BEWARE THE CON ARTISTS—

Be aware of service station attendants who are simply out to take you. A great number of gas stations across the country are operated by con artists. They will get a few dollars out of you if you give them half a chance. Be alert and informed of what they are capable of doing.

Never leave the car unattended at a strange gas station. If you have to go to the restroom and there are two or more of you traveling, leave someone with the car all the time. If you are alone, go to the restroom *after* the car is completely serviced.

If the attendant checks under the hood, amble around so you can keep an eye on his hands. The oldest trick in the world is not shoving the dipstick all the way in the engine. Naturally, when he pulls it back the engine is a quart low on oil. That trick costs you one quart of oil which you don't need.

If you're not watching, a good shove or tug on a heater or radiator hose can rupture it so it will leak. That's all the con needs to start his speech about how bad the hose is and how dangerous it is to go back out on the road with it in that condition. The same thing can be done

with a fan or accessory belt. A knife can even be brought into play here to put a healthy size nick in a belt. There's plenty of time to do a lot if you are away from the car.

If you have a tendency to have faith in people, you're just that much easier to victimize. In a garage or even in a service station where you are having something minor done they can easily get you for a tire, battery or set of shocks once they set out to do just that. For as they work, they "keep finding things" wrong with your car.

Women are particularly vulnerable because attendants assume their knowledge of cars is limited. A fast-talking con artist can put fear into a woman—convincing her the car she is driving is jeopardizing her life. A little knowledge and a cool head can go a long way towards saving the day and many dollars.

"I THINK WE'RE LOST."

When traveling an unfamiliar freeway or highway in a metropolitan area, it's easy to find yourself in the wrong lane for a turn you need to take. When traffic is normal, changing lanes at the last second can be difficult and sometimes dangerous. When traffic is heavy, the same maneuver can lead to disaster. Thus on strange ground it is very important to look far ahead for any type of sign which will lead you to a lane change or turn.

When traveling to an unfamiliar area, get in the habit of spending some time studying a map of the area. Sure, you know where you are going and roughly how to get there, but do you know the highway numbers—and are they state, county or federal roads? When making that decision to turn or go straight, knowing the highway numbers can save the day. You needn't try to remember every highway number you'll encounter on the trip—just the two or three that have a direct bearing on your travel plans of the day.

Despite your protestations about the "idiot that laid all of this out," highway and traffic engineers are quite systematic, thoughtful and consistent with their markings. The sad fact of the matter is that most of us don't read signs—we glance at them. Patience does not come easy (if ever) to the fast-moving American. How many times have you pulled into a service station to ask directions only to be told you're on the right road.... "just keep on.... you can't miss it."

Even knowing that you should be generally traveling west instead of south can help a great deal when coupled with reading (not glancing) at state and federal highway markers—because many of them indicate the direction being traveled in addition to the number of the highway. Our interstate highway system has grown so quickly that only a small percentage of drivers even know that odd-numbered interstate highways are north/south routes and east/west routes are always even-numbered.

Most all exits and turns on modern highways are indicated twice—Bell Road Exit 1 Mile, Bell Road Exit 1/4 Mile. Be alert. Look for the first sign—catching it makes the second sign easy to spot. Be alert; read—don't glance.

Missing a critical exit or turn in a strange city can be a real hassle—double

How often have you seen a confused tourist change his mind at a poorly-marked intersection like this while zigging back and forth between lanes? If he makes the wrong decision on a limited-access highway, he can drive miles before he untangles the maze and gets headed in the right direction again. Study your map at rest stops and ask a passenger to keep track of where you are. Then you can guess right—when you have to guess.

so if you are towing a trailer or driving a truck. Three minutes spent with a map before the trip can save 30 minutes later on, plus a lot of headaches—there's no way to calculate the amount of gas you'll save.

NO IMMOLATION PLEASE

The fear of running out of gasoline has so possessed a large number of drivers that they have resorted to carrying auxiliary supplies with them on trips. Until you have seen a car burn or explode because of gasoline there is no way to emphasize the seriousness of the situation. Carrying fuel in cans—of any kind—inside any vehicle is extremely dangerous. Vapors, fumes, leakage, a spark, a collision—all factors in a lethal combination from which there is little escape. Extra gasoline should be carried in an approved and properly installed auxiliary gas tank which augments or replaces the stock fuel tank in your vehicle. If you can't do this, leave the extra fuel behind.

AN AFTERNOON AGAINST THE SUN

Driving into the late afternoon sun can be a pain in the brain (and to the eyeballs). This is the time when motorists are hurrying home from work in heavy traffic—a prime time for mishaps. All of a sudden you crest a hill and find the sun blinding you. Cars ahead are only a silhouette as you squint and try to maintain direction. Many motorists panic in this situation and either lift the throttle completely or set the brake. If you give yourself enough room, the situation normally won't call for making harsh moves. When driving in any adverse situation, leave a little more space than you normally would between you and the car ahead. Be alert. Know what the car ahead of you is doing—even if the car is a silhouette and the tail lights are in a deep shadow.

If you must carry extra gasoline, carry it outside of your vehicle. But before attaching fuel containers, check to be sure it is legal. In some states you can get a ticket.

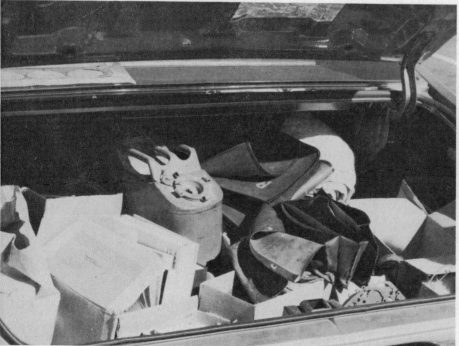

This is very dangerous. Carrying gasoline around in the trunk of your car is probably more hazardous than carrying a couple of sticks of dynamite in the same location. Certainly the gasoline is a lot easier to touch off.

Be alert to the terrain and type of road you're on. If it's a curvy road, take in consideration that you may get a reprieve momentarily from the sun hitting you in the eyes, and all of a sudden be faced with it again as you swing around the turn.

If another driver cuts in ahead of you because you've allowed yourself a safety margin, feel free to mutter under your breath—and then carefully build yourself another margin of safety. Instead of a life being lost under these hazardous and annoying conditions, only a few seconds will be lost.

Confidence and patience along with defensive driving is the best way to save on fuel and the safest way to drive. Even at a time when economy is an important factor, it must take a back seat to safety. There are no savings in a mishap.

When you are behind the wheel, be a full-time driver!

A SPEEDING TICKET IS NO ECONOMY

Highway patrol surveillance all over the country increased when the 55 MPH speed limit went into effect. When there's little traffic, your chances of getting a speed ticket are greater than ever. I had an experience which may help those who insist on speeding.

While driving in California, a highway patrolman watched me for about 80 miles! The game started early in the evening about dusk and went on into darkness before the surveillance ended. Obviously, the officer didn't have much traffic to work and I happened to be in his sights. The cat-and-mouse game began just out of Indio, California. On three different occasions during the next hour and a half, the patrolman pulled off the highway and out of sight—on an overpass or behind a knoll. Gut feel told me I wasn't alone out there in the desert. Mile after mile would go by and then a pair of headlights would appear far behind me. For several miles they would close ever so slowly. Needless to say I was very careful in observing the speed limit during these unusual efforts to find a taker for a ticket. As an extra precaution, I dropped a mile or two below speed limit—just in case. After five miles or so of this, the patrolman would pass me and rather rapidly go on out of sight. When I finally got to the next town, I pulled in for a coffee and monotony break. There, sitting at the counter was the highway patrolman who had "protected" me for that long distance.

In more than two-million miles of driving I have learned that most law-enforcement officers respect an alert driver and will not ticket you unless you are way out of line. My respect is high for the majority of highway patrolmen. The point of telling the previous story is that with fewer cars on the road and the same number of lawmen, there is a very good chance you're going to be looked at much closer. After all it's the officer's job to watch traffic.

On another occasion, I saw a highway patrolman traveling in the opposite direction on a four-lane highway with a wide center divider. I was fifty'ing along fairly carefully in an inconspicious compact stationwagon. John Law had that straight-ahead-going-someplace-look as I watched him out of the corner of my eye while I put on my not-aware-of-him look. This was wide-open country and there wasn't another car in sight for miles in either direction so I felt I hadn't seen the last of the all-business lawman.

About the time the patrol car appeared no bigger than the head of a pin in my rear view mirror, I saw him U-turn across the median and watched the car grow progressively larger as he came up behind me. He followed for two or three miles at a distance of 1/3 to 1/2 mile. He then moved to within about 1/4 mile and paced me some more. I had the speedometer glued on 54 MPH because this guy was working too hard to suit me. This was at the height of the energy crunch and customers for his tickets to the policeman's ball were hard to find.

Approximately another mile down the road the flashing lights beckoned me to pull over. A very business like officer informed me I was going 58 MPH. He said he was letting me off with a warning. On that note I could see no need of mentioning the 54 MPH I was religiously holding to—just a quick, "Thanks a bunch. I'll watch it more closely."

The difference in our speedometers caused me concern so I dug through my briefcase and pulled out a stopwatch in preparation for a time/distance check which I knew was located further up the same road. My speedometer was in error. The first mile took 62.9 seconds so the speed was 57.2 MPH. In the second mile I slowed to an indicated 53 MPH and the

This chap intends to carry gasoline in the same compartment with people. A wienie roast is lots of fun if you are not the weinie.

These citizens are normally helpful and fair-minded. Occasionally you'll spot one who seems to think he can give you a ticket if he watches you long enough or plays peek-a-boo for the next fifty miles. If you find yourself the victim of this game, do a lot of peeking.

time came very close to 65.5 which is correct for 55 MPH over a measured mile. So when I had been running 55 MPH (indicated) I really was running about 57 MPH—and the patrolman said I was running 58. I'll have to give him credit for being close. So speedometers can get you into a little trouble if the law is going to hold you to within 2 to 3 miles an hour.

TRAILER TOWING

The effect a trailer has on gas mileage varies greatly from one towing application to another. A large truck or motor home towing a small or even medium-size trailer might lose less than one mile per gallon on a long trip. On the other hand a small sedan or compact car pulling a trailer with a large frontal area might wind up losing nearly half its gas mileage—due to increased air drag.

Substantial gains can be made in fuel economy when a trailer is chosen with good aerodynamics. Tests run by U-Haul have shown that drag can be reduced 30% by rounding all the corners of a travel trailer on a 6-inch radius.

Simply stated, if you tow any kind of trailer that presents more frontal area than the towing vehicle, you can expect very poor fuel economy. Towing with a vehicle larger than the trailer shows less reduction in economy at steady highway speed.

QUICKIE TRIP CHECK LIST

1. Accessory belts—tension, fraying, cuts
2. Battery—water level, terminals clean and tight
3. Brake fluid—level
4. Heater hoses—abrasions, cracking, seepage
5. Oil—level
6. Power steering—fluid level
7. Radiator—water level
8. Radiator hoses—any evidence of imminent failure?
9. Lights—bright, dim, directional and stop
10. Tires—inflation, cuts. Check the spare! Got the jack?
11. Transmission—fluid level

With the exception of the brake lights, one person can perform the above check of most any car or light truck in about 20 minutes. Back the car up near the garage door at night if you want to check brake lights without help.

A "slice of life" at a roadside rest stop. The motorhome in the center with some attempt at streamlining should get better highway mileage than the squared-off model at left. The motorhome in the center can tow the car behind it without a large penalty in highway fuel economy because the car does not add frontal area to the towing vehicle. The small car at right is paying a high price in frontal area and gas mileage due to that box on top. A small trailer could be more economical on a long trip. The wrecker out there is still looking for a customer.

IS PREMIUM BETTER THAN REGULAR?

The idea has gotten around that premium or high-test gasoline is somehow superior to plain old regular, and some people buy it just to be nice to their cars or in hope that it will prolong engine life. Premium gasoline is certainly better for engines that require it. In a car that doesn't need it, it's a waste of your money.

Best practice is to buy the grade of gasoline specified in your owner's manual—unless you have made engine modifications to change the engine. Some older engines get "carboned up" on the inside and won't run anymore on regular without pinging and continuing to run after you have turned off the switch. In some cases a change to premium gasoline will get a few more months or years of service before taking it to a mechanic.

4 SAVING IS NOT SPENDING SO MUCH

There are several approaches to economy. One is to buy something and try to get the most service or use out of it. Another is to pay less for what you buy. That's what this chapter is about. Saving is not spending so much!

After you buy your car, what two things are the greatest expense of driving it? One of them is gasoline and you said that instantly, didn't you? The other may be insurance, depending on where you live and what you drive. Bet you didn't think of that! How much *do* you pay, by the way?

WHICH GAS?

There are two ways to buy gasoline for the family car. The first is to visit the brand-name station on the corner, flash your credit card and order. Once in a while you'll find an attendant willing to check the oil, water, and battery.

There's another way though—time honored by some of us wanting to stretch our dollar bills to the limit. Drive past the name-brand station and go to the pump-it-yourself place. The station probably sells a brand you never heard of; you'll have to pump your own and they probably won't take checks or credit cards. This is strictly do-it-yourself: No one checks anything and you're in the wrong establishment if you need a flat fixed. Why buy gas there? Well, if you compare, you'll probably find gasoline is a few cents a gallon less than at the name-brand stations in the same area.

You can save a buck by filling up yourself. Maybe you knew all this but have wondered about the quality of gas at these stations. They buy gasoline from all major refineries—you might not know if you are getting Shell, Texaco, Exxon or some other major brand, but how much do you care if the price is right and the gas is good?

Should you buy regular or premium gasoline? If your car is tuned for regular you won't get any better gas mileage by burning premium. If the car is tuned for premium and you run regular, you won't gain or lose gas mileage, but you run the danger of having the engine detonate, also called knocking and pinging. Prolonged detonation will ruin an engine. Some cars can be tuned to burn regular gasoline by a simple adjustment of the ignition timing. On other engines, this won't work. If you are in doubt on this, consult a sharp mechanic.

NO-LEAD GASOLINE

If you have a 1975 or later model with a catalytic muffler, you are stuck with using no-lead gasoline. However it is also sold at pump-it-yourself stations where you can make a saving of a few pennies per gallon.

Old timers will remember the days when no-lead gasoline was the standard product and you had to pay extra to get gasoline with the lead compound added. Today you have to pay extra to get them to leave the lead out—which of course is progress.

Anyway, the combination of a special catalytic converter in the exhaust system and the no-lead gas which is required for that system is supposed to improve gas mileage by 10 or 15% and tests on 1975 models indicate the claim is supportable by facts in many makes.

THEY'RE INDEPENDENT, YOU'RE INDEPENDENT

A word of warning. If you patronize the fill-it-yourself gas stations, you must be willing to assume all of the responsibility for normal filling station services that you won't get anymore. Don't learn this the hard way.

Brand-name stations justify their higher price for gasoline by reminding us that they do other things: Clean your windshield, check engine oil, check battery water level, look over the engine compartment for visible signs of trouble, add air to low tires and so forth. It's true that much of this is lip service rather than real service and the willingness of the attendant to give you the full treatment depends on how busy the station is, whether the boss is watching or not, and

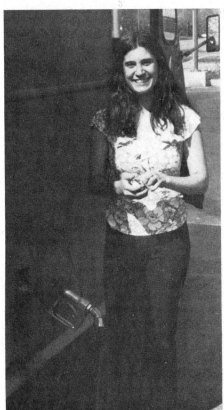

Pumping your own gas at self-serve stations is no big deal. Instructions on the pump tell you how. If you haven't done it before, most station attendants will be glad to show you how. Don't smoke while pumping gas. Don't forget to put the gas cap back on. Lay it on the driver's seat and it's pretty hard not to notice it when you get ready to drive away.

sometimes how insistent you are.

Nevertheless, if you buy gas at brand-name stations you will occasionally get everything checked and that's a heck of a lot better than *never* checking it.

If you just fill up at the discount pumps and drive away, months can go by without *anybody* checking your oil or battery *or anything!* One day it won't start, or it makes expensive noises at you.

Most people can figure out how to operate a gas pump and air hose even if they never done it before. And most have some practical experience under the hood. The first part of this chapter assumes average know-how on your part.

But if you are just getting interested in saving bucks by doing your own routine maintenance, you may never have lifted a hood. Don't laugh, but if you've never done it, the thing may defeat you on your first try because you won't know about two latches in two different locations, or whatever the trick is on the car you drive.

At the end of this chapter is a special section for beginners. It will teach you at least as much as the average gas-pump attendant knows and probably give you less bum info than you get from that greasy-faced cherub who tells you non-detergent oil is best. It isn't.

I will forthrightly demonstrate such matters as raising the hood, changing the engine oil, checking and adding transmission fluid, peeking in the battery and lots of other stuff. You can read it in the privacy of your own bedroom and nobody will know.

ECONOMY INSURANCE?

Basically, there are seven or more kinds of insurance affecting your car. Or more correctly, affecting you when your car is involved in some unfortunate incident. Common types of insurance are—liability, collision, property damage, comprehensive, medical payments, uninsured motorist, and towing or road service. If the vehicle is damaged in a wreck, *collision insurance* hopefully pays the repair bill. If you are hurt or someone else is hurt, *liability insurance* covers that portion of the damages. If the driver who hit you has no insurance, then your *uninsured motorist* coverage takes over. If your car is stolen or catches on fire, the *comprehensive policy* comes into play. The *medical pay-*

The last gas shortage caused a lot of people to stock up before the hoarders got it all. If you start an enterprise like this in your back yard, you are violating probably a dozen laws including one called *common sense.*

Sometimes a little hurrying or inattention causes a tragic ending to a pleasant drive. Be careful!

ments portion of a policy is written to pay off regardless of who is at fault.

Liability, collision and uninsured motorist coverage are normally sold as a package. If your car is an old one and you place little value on it, you might choose not to have collision coverage.

Here's a bit of philosophy you may not have considered. It concerns how you manage the risks in your life. Every part of living has risks. According to some of the cheery news events lately, even breathing the air or drinking the water can kill you in some parts of our fair land.

Anyway, you have to decide what risks you face and what you are going to do about them. Some risks have an end result that you urgently wish to avoid. Therefore you avoid exposing yourself to that hazard. This may cause you to decide to leave airplanes only at the regular passenger loading points listed in the schedule. If you have a family, you may wish to protect them against the consequences of your sudden end, so you buy life insurance.

There's an important distinction. Life insurance doesn't keep you from getting killed, it only makes things easier for your survivors. To keep from getting killed you have to be careful—sometimes lucky too. If that seems obvious, you are a straight thinker.

Another risk with a very bad end result is a lawsuit over a traffic accident. If you lose, the jury will probably give away your total earnings for about half of your lifetime. Most people are very happy to pay the insurance company to assume that risk, and it is required by law in a lot of places.

In the car-insurance field, most of the risks are optional. You can take the risk yourself or you can pay the insurance company to take the risk for you. Now it becomes difficult to decide what to do.

Take towing and road service for example. The insurance salesman points out that the coverage is very cheap—only two or three bucks per year. You can easily spend 5 to 50 dollars if you break down on the highway or your driveway and need towing or other assistance. You have to make an individual decision based on several things.

If 5 or 50 bucks is usually not in your wallet but you can scrape up 3 dollars once a year for insurance, it may help your peace of mind to buy it. If you can afford a tow bill but you haven't had one in the last 20 years, then you should consider saving the insurance fee.

Now what? If you can't fix it yourself, you will need towing or road service. Car insurance coverage can pay all or part of the cost. Do you have it? Do you need it?

Towing insurance is penny-ante compared to collision and I only talked about it to prepare you for the biggie. Do you really want to pay the tab for collision insurance on your car? If you don't, and you get in a crash, you have to collect from the other driver or his insurance company or pay it yourself. That means you must not be at fault in the accident. That also means the other driver must be at fault and have some money or some insurance so you can collect.

How do you analyze a risk like that? For one single individual whose name is yours, you can't. For a large group of people—or for the average person in that group—it is possible to figure the odds statistically. That's what the insurance company does and how they arrive at their rates.

They figure the odds on your getting smashed and make you a bet about it. They choose odds that are in their favor—or at least that they think are in their favor.

You can figure the odds on that bet just by comparing what you pay for collision coverage to the value of your automobile. If insurance costs $100 a year and the value of your car is only $500, then the insurance company thinks you have 1 chance in 5 of getting smashed this year. Actually the insurance company thinks your chance of smashing is less than that because they intend to make money after paying everybody off.

If you think your chance is less than that, then maybe you shouldn't take the collision coverage. But remember, either way you are *gambling*. The rule in all kinds of gambling whether at Las Vegas or Wall Street is: *Don't gamble money you can't afford to lose.*

If you couldn't come up with the money to replace your car then you should probably carry collision insurance, but don't jack up the value of the car because they'll only pay what it's actually worth. If it's an old car, not worth much anyway, and you could afford to replace it, then consider assuming the risk yourself.

Talk it over with your insurance man—but keep in mind the fact that he sells insurance for a living.

Nobody can tell you how to get the best deal on insurance because in many ways it's like selecting a husband or wife—personal preferences and decisions are involved. Some general guidelines can save you money.

BUY DIRECT

Insurance companies offer insurance in several ways. Some of the biggest have salesmen or agents who sell only that brand. They usually have an efficient department to process claims in volume and give you the same kind of courtesy a new boy gets when he reports to the Army. But it's hard to beat their price. That's why they are the biggest.

Other companies sell through independent agents who often advertise that they can serve you better because they can place your policy with any of several different insurers. It depends on what *service* means. Often the price is higher than direct buying but some people find it advantageous to put all of their insurance with one independent agent and gain some leverage with the agent. It's possible that he can save you some money in your total cost of insurance if you have several different policies and things to insure. He can help you avoid both unnecessary duplication and errors of omission. But he can rarely save you any money on car insurance alone. If the agent starts moving your insurance to a different company each year, be wary.

CHECK OUT THE SPECIAL DEALS

Ask your friends about their insurance. You may find your next door neighbor buys his car insurance by mail from a place in Virginia, saves a hundred bucks a year and is perfectly happy with the arrangement.

Some insurance companies sell only to special groups such as government employees, union members, or employees of a particular company. All states have an insurance commissioner who monitors activities of insurance companies doing business in the state. Usually you can collect from any company authorized to sell you a policy, but collecting may not be simple or easy. It may involve running around town to get estimates on your repair and then settling the deal by mail or long-distance phone.

If you have two cars and can get by a while with only one, it could be worth it to save the bucks.

CHECK OUT THE SERVICE

Insurance companies of the same type, such as those who sell direct, usually have very similar rates. In that case, it pays to check their general reputation with their own customers. Ask around among your friends. You will find some who think their insuror is fast, fair and friendly. Others will report a terrible hassle whenever they have a claim. If the cost is about the same, why not go with the good guys?

CHECK OUT THE POLICY PROVISIONS

Every benefit in every policy has some limits and conditions and you should know exactly what you are buying. It's pretty sad to crash into a telephone pole and *then* discover you have insurance for that if it happened on Tuesday afternoon.

This is a very good reason not to buy insurance over the telephone. You can do it in many cases, but how sure are you of what you bought? When the policy comes in the mail, you know you're gonna toss it in a drawer and you may *never* know exactly what your coverage is. Maybe you will find out when it's too late.

Many companies now give you a short printed summary of what the policy says and these are honest but only summaries. Read the summary. Then read the exact language of the policy—at least when you first start buying auto insurance. Meet the salesman *face to face* and ask questions until you are sure you understand what the policy provisions are. Educating the buyer is part of his job and if he doesn't think so, it's easy to find another salesman who does.

If an insurance salesman says he can cut your car insurance cost by 30%, let it register with you that he may be cutting your insurance protection by a like amount. Use all the caution you would in selecting a doctor for your child. Talk to several people who have been involved with their insurance companies—what good does it do to talk insurance to someone who has never had a wreck?

GET THE "SAFE-DRIVER" DISCOUNT

The economy in car insurance comes mainly from your driving record. If you have run up a string of tickets for moving traffic violations, been arrested for drunk driving or have been involved in a wreck or two in the previous three-year period you can expect both liability and collision insurance to be very high. There is simply no way around this.

Remember too, the harder a company looks at your record, the better off you are if the record is a good one. If your driving record is so bad an insurance company cancels the policy, you are faced with expensive insurance for a long time into the future. Never, EVER, make a false statement when applying for insurance or filing a claim. If any falsehood is ever discovered by the company by a routine spot check or through investigation of a major accident, the company has a legal right to cancel the policy and not pay the claim! Very poor economy on your part!

You may have heard you can post a bond instead of having a liability policy. This is true but liability performance bonds are expensive. If state law requires a $15,000/$30,000 liability policy or the same amount in a performance bond, you'll have to shell out about $1500 a year just to keep the bond in effect. Makes insurance a pretty attractive deal after all, doesn't it?

Consider the following when shopping for insurance:

1. You definitely should have liability insurance—doing without is false economy and may be illegal.

2. Deal with a reputable agent.

3. If you insure more than one vehicle with a company, you normally get a discount.

4. Before buying collision insurance on an older car, ask yourself if it's economically practical.

5. Don't buy car insurance (or any other kind!) over the phone. Sit down with the agent and go over what you really expect the policy to do for you. Read the policy and make sure it says what you think it says.

6. Driver age and driving record are the key factors in determining what you pay for liability insurance.

7. Does the liability insurance cover you in another car, or another driver in your car? How about minor drivers?

8. Although you may think car-rental insurance is highway robbery—it will save you untold amounts of time and money if you are ever in a rental car and have a

wreck. If you travel a lot and use rent cars frequently, ask your insurance agent's advice on this.

9. Don't be too quick to get a performance bond in lieu of liability insurance—it is expensive.

10. If you have a poor driving record, don't think you can beat it by lying to a new insurance agent—they check your driving record with the appropriate state agencies as a matter of routine.

11. Does the policy have a non-cancellation clause?

12. Some companies offer monthly payments—but you pay extra for this. The lowest premiums are based on annual payments.

13. You should carry proof of insurance with you at all times.

14. Don't figure the insurance company will make required accident reports for you. They'll usually supply the forms and instructions.

15. Your U.S. insurance probably does not cover you in foreign countries, including Mexico and Canada.

TIRES

You can save a lot of money on tires—or you can spend a lot. Sometimes you can spend a lot and save a lot in the same deal. The first rule is to start shopping before you are in dire need. In other words, don't run tires until you have a blowout; then discover to your horror and absolute amazement the remaining "good" tires are bald. If you must leave on an 800-mile trip tomorrow, the amount of time you have to shop for tires is limited. So is the possibility of catching a tire sale—or maybe even the chance of getting exactly the tire you want or need.

Let's say you are flogging a '65 Chevelle back and forth to work and you are trying to keep close control on dollars spent on the vehicle. After looking the situation over one morning, you come to the conclusion you really don't want to drive to work on those tires more than another three months. Several choices are open to you.

Check the local paper for sales on tires. When you fill up with gas, ask the attendant what he has in the way of a set of tires for your car. Check a high-volume tire store to see if they have a set of used tires from a brand-new car. The store cannot legally sell the tires for new ones—but for your purpose they may be perfect. There are also tires that for one reason or another did not pass the manufacturer's strict grading test. Maybe the white sidewall is blemished; maybe the tire is tubeless and has tiny pin holes in the carcass and won't hold air. The price comes down accordingly. In the case of the blemished sidewall, turn the whitewall in and run the tires as blackwalls. In the case of the leaking tubeless tires; buy a tube.

As you can see, you can spend less for tires. The key to success is to plan ahead and do some shopping. If you are forced to buy tires on short notice, you can count on paying a premium price—and sometimes you won't even get a premium tire.

You should know roughly what kind of tire you want before you ever start shopping. Make an honest assessment of what the tires are to be used on and how they are to be used. If a car or truck is to be used for your work it may put a far different perspective on tire buying than if you plan to use the car mostly on city streets to run errands or commute a few miles daily. Buying the most expensive radial-ply tire for a "grocery getter" makes no more sense than putting the cheapest tire you can find on the car you drive 800 miles a week on the freeway. To make an intelligent choice on the type of tire to buy, listen to a tire salesman—but make your own decision. First be armed with an understanding of current tire technology and the terms that are used.

Three types of tire construction are available today: *Bias, belted-bias,* and *radial.* Bias construction tells how the plies or layers of fabric are placed around the tire. The fabric cord is "on the bias," or placed on an angle of 30 or 40 degrees to the centerline of the tread. The bias-ply tire is the oldest of the three types listed and will probably be with us for a long time to come. Strength is gained simply and effectively by adding more plies. As strength is added in this manner, the bias tire gets stiffer and gives a harsher ride.

Most tires sold in the U.S. today are of bias construction and although their popu-

Buying tires isn't a steady leak out of your wallet like buying gasoline. Tire purchases take a bigger bite less often. Nevertheless, an intelligent choice and careful shopping can save you a significant amount of money.

If you don't see tires advertised at sale prices—wait a few minutes. Few commodities are more often "on sale" than tires.

larity has been severely reduced in the past few years, you can count on bias-ply tires to be around for a long time to come—strength, versatility of design and comparatively low manufacturing costs are all on the side of the bias-ply tire.

Historically, the radial-ply tire came next in our lineup of the three basic types. On a radial tire, the cords or heavy threads of the fabric go across the tire from bead to bead making a 90° angle with the centerline of the tread. One or more "belts" of flat fabric or steel mesh are laid under the tread to stabilize it. This leads to vastly superior traction under all conditions. A radial is long wearing and smooth riding. They are slightly more vulnerable to sidewall damage than the other two types of tires. In terms of tractive effort—how good they grip the road—the radial tire is simply great.

The bias-belted tire is an attempt to bridge the gap between radial and regular bias construction. Bias-belted tires are a bias-ply tire with a preformed "belt" of cording placed under the tread. The sidewall remains totally bias, but the tread area gets tougher, and the tire is smoother running and longer wearing than the ordinary bias construction. Among car enthusiasts, this is the current "in" tire—due mainly to the wide "low-profile" look.

If you don't understand all you see printed on the side of the tire these days, don't feel like the Lone Stranger. Tire nomenclature has always been somewhat confusing to the layman and appears to be getting more so by the day. Actually, things are getting simpler—but there are a few lines to learn before trotting off to the local tire emporium.

The layers of fabric used in tire construction are called plies. Plies are used in even numbers. The more plies, the stronger and stiffer the tire becomes. Plies may be constructed of rayon, nylon or polyester. Rayon is now used for the plies on inexpensive tires, but it is also used on premium radials in conjunction with a steel belt. Nylon is probably the most commonly used material for plies today. It is quite strong. To some degree, most nylon will take a set when parked overnight or for any extended period with the weight of the car concentrated on one small area of the tread. The tires will make a thumping noise for a mile or so when first driven. This actually sounds and feels like there is a flat spot on the tire. As a matter of fact, there is and this is normal.

Polyester body plies and a fiberglass belt under the tread are fast becoming a popular combination for belted-bias construction. Ply material is constantly being upgraded by the tire manufacturers. It is almost impossible to get a "bad" tire. The trick is to get the tire you and your car need that will serve you best.

The number of plies used in the construction of a tire was a pretty simple matter until the mid-1960's when tire makers started saying they could make a two-ply construction at four-ply rating—the rating being a measurement of strength. This was about the time the Federal government was taking a little closer look at the automobile industry anyway, and the tire got shoved under the fed's microscope. One might guess that all which was observed did not match marketing claims. At any rate, we now have something called Load Ranges. A new tire today will be imprinted Load Range A. This means the tire will hold as much load as a two-ply tire. Range B is four-ply and so on. Because a lot of the buyers are still scratching their heads over this one, companies are trying to help out by printing the Load Range *and* the number of plies along with the old ply rating. Thus you get **Load Range B 2 ply/ 4 ply rating**. A tire shop can show you a chart which gives all of this in pounds.

If you see numbers on the tire sidewall like 6.70-15, you can tell the size of the tire. the **15** (14, 16 or whatever) refers to the diameter of the rim that the tire will fit. The number **6.70** refers to the width of the tire in inches. If the entire number (6.70-15) is wholly numerical, the tire is of bias construction.

If the number contains a letter such as **G70-15** or **J60-15**, then the tire may be bias-belted construction. On tires with this type of designation, the letter—A through L—refers to the width of the tire. A is the smallest, L is the widest. The next two numbers—**60** or **70** in the examples given above—refer to the aspect ratio of the tire. Aspect ratio is a fancy term which means you divide the height of the tire by the width. Thus the **70** series tires are 70% as tall as they are wide. The last number, such as **15**, is the rim size again.

A radial-ply tire will always have the letter R somewhere in the size information. It might be something like **LR60-15**, which means an L-60-15 radial.

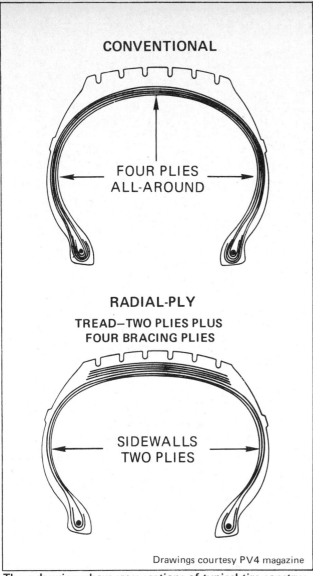

These drawings show cross-sections of typical tire construction methods. Note the reduced number of sidewall plies in the radial tire but the added belts beneath the tread.

Tires today carry full consumer information imprinted on the tire itself. Read and understand before you buy.

Naturally, this is too simple to be consistent, so you'll find designations like R225-15. This also means radial construction but the **225** refers to the width of the tire in millimeters. As far as I know, this only shows up on tires made outside the United States.

While you are standing on your head trying to figure out what kind of tires are on that Chevelle, you'll probably run across something that looks like **DOT JAL 5 AKMD123**. The **DOT** refers to the Department of Transportation; everything that follows is their method of identifying who built the tire, how it is rated and when it was built. When it was built can be determined by the last three digits. In the above example, the **123** is the key to the date of manufacture. The **12** means the 12th week of the year and the **3** means 1973. One last item concerning the **DOT** designation. If you see **DOT R**, this does not mean the tire is of radial construction. If the R is adjacent to the DOT, the tire is a recap. A lot of changes have occurred in tires and tire designations in the past few years.

Tires come in tube and tubeless construction. If they are tubeless, this will be printed on the sidewall. If they are of tube type, then chances are no mention will be made of the fact.

Aspect ratio is the height of the tire divided by the width. A "70 series" tire is 70 percent as tall as it is wide.

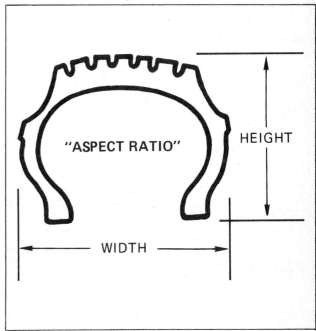

If all of this sounds like a lot of writing on the side of a tire, the bright side of the picture is that you will have plenty to read while changing a flat.

At one time, if you were looking for a low-cost tire, a recapped tire—old carcass, new tread—was the most economical way for a short-hauler to keep the rims off the road. They weren't always best for long fast trips because heat and speed did many of them in.

Nowadays there are federal standards for retreads. Quality control is much better, but prices have gone up. In some areas it is difficult to find a retread which has much if any price advantage over a new tire. Try to sell a set of good tire carcasses to a recapper as you could five years ago. He might not take them off your hands for free.

The tire to buy is based on what you use the car for and how long you plan to keep it. Generally speaking, radials give the best performance, best gas mileage, and will wear longer than a tire of any other construction. Radials cost more than other tires when you buy them, but are the most economical in the long run—IF THE RUN IS LONG ENOUGH. Bias-belted tires are the compromise. They won't wear as long as a radial but they cost less and give excellent performance. Bias tires are probably your best bet if you don't plan to keep the car for a long period of time or are running a light or medium-size car without carrying heavy loads or running at constant high speed.

You get a steady hammering by experts in the popular press and government to persuade you that there is a guaranteed gas mileage improvement in a set of radial tires. That's true to a degree, but it isn't really that simple. If you think about it for a moment, you will conclude that if radial tires give better fuel economy then they must have lower rolling resistance than other types. They generally do, because of the stabilizing belt under the tread.

Now think about low-speed and high-speed driving. At low speeds, wind resistance is also low and it is rolling resistance more than anything else that uses up energy to move the car along the road. At low speeds the radial tire may give an improvement in economy.

There's no guarantee that you'll get better mileage because I have run some tests on a Vega where no improvement was gained by using radials.

But at high speeds on the expressway, it's wind resistance that does it to you. The higher the speed—the more dramatically the wind resistance increases—and the lower your gas mileage becomes. A radial tire can't do much about the wind resistance of the whole car, so you can't expect much help from that source. If you do a lot of driving with one eye on the road and the other searching the bushes for the law, buy radial tires for sure but don't expect them to save you a bunch of money. You speeder!

Armed with all of this information, there are only a few more things to keep in mind before dealing with the tire salesman. In price comparison, don't overlook sales tax and the special federal excise tax on tires being used on public highways. Off-road tires are exempt from this tax. The reason you should ask about the taxes is so you'll be comparing apples to apples when you go shopping. The same for mounting the tires and for balancing. In many stores the price of mounting and balancing is included in the price of the tire. That's fine until you start comparing prices from a store where these services are not included in the price.

If you pay list price for tires, don't tell anybody. Every tire outlet has frequent sales. The big merchandisers have sales so regularly you could set your calendar by them. As mentioned earlier, do your shopping while you still have a month or two left before you actually have to put on new skins.

When you have decided, write down the size, type of construction, load rating and price. Don't buy yet. Some people have hangups about tire brands and if you think one is significantly better, I won't try to change your mind. But remember, with the feds policing the industry, all tires with the same specifications have to meet the same minimum standards. "El Supremo" brand may be just as good as the name brand you have been using—in fact it may be built in the same factory.

Anyway, if you shop for specifications rather than brand and wait for somebody to have a sale, you'll save some money.

TIRE BALANCING

New tires should be balanced after they have several hundred miles on them. Most tire shops will balance the tires at the time they are sold and mounted on the car. That's fine; but the balance job that is really important is the one done after

These drawings show three common tire constructions. Bias-ply means that the cords in the tire fabric run at an angle across the tread area. Radial plies have cords which run straight across the tread area—from one side of the tire to the other. Radial ply tires always use strengthening belts under the tread.

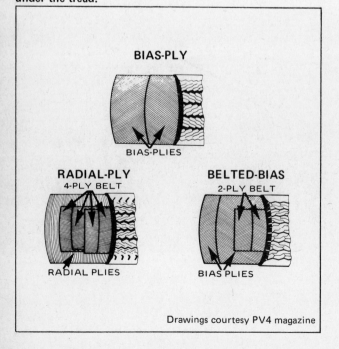

Drawings courtesy PV4 magazine

Busy tire shops such as this one operated by Dick Cepek in California use precision spin-balancers to balance both tire and wheel. This is the best type of balance job.

Maintaining wheel alignment is a problem. It costs money to do it and can cost more if you don't. Tire wear guarantees may not apply to tires badly worn due to wheel misalignment although some dealers adjust anyway to maintain customer good will.

several hundred miles of driving. Tell the tire salesman that is what you want. Work out your own deal on this. Get a receipt for a free balance job later.

There are balance jobs and there are balance jobs. There is static balancing and there is dynamic balancing. The first one you don't want. This is where they lay the tire on its side and then try to put a little bubble in a circle. Dynamic balancing is also known as spin-balancing. Currently this is the only real way to get a tire balanced correctly. Depending on the tire, the vehicle and the type of driving involved, a good balance job can add several thousand miles to the useful life of a tire. And, you enjoy a smoother ride in the bargain.

WHEEL ALIGNMENT

Today many tire stores have the necessary hardware to perform a simple front-end adjustment. Many shops do this as a routine when selling a new set of tires. This is a worthwhile service that helps get maximum wear from the front tires.

Alignment gets jolted out of adjustment when you hit a curb or hole in the road. Some cars can keep perfect alignment for years, others can go through several legitimate front-end alignments a year if the roads are rough or the driver careless.

Legitimate front-end alignment? You mean there's another kind? Uh huh. The front-end alignment you get for free at the tire shop is pretty slick—especially slick if you look around and see that they just sell tires. A shop that also sells ball joints, tie rods, springs and shocks can find all sorts of things wrong with a front end. Many people fall for this—hook, line and sinker—because the guy selling the merchandise has a sales pitch that is hard to refute: The car is unsafe to drive. Refute anyway. A tie rod might be bent, a spring might need to be replaced—maybe, maybe not. This doesn't always make the car unsafe.

Running with your front wheels out of alignment is poor economy because it will chew up the tread in a hurry. Buying a lot of other stuff you don't need is also poor economy. Get your alignment at a reputable shop and get a second opinion on anything else they want to sell you.

SNOW TIRES

The economical way of contending with snow tires is to buy a set of unbent wheels at a wrecking yard and have the snow tires mounted. Sure, you'll be out the purchase price of a couple of wheels, but you'll have only one mounting and one balancing fee per wheel instead of two bills for this every winter and spring.

Additionally, it doesn't make much sense to drive through a snow storm on regular tires to a service station to have the snow tires mounted, does it?

With snow tires already mounted on wheels, it's a simple matter to switch back and forth between snow and regular tires as you choose.

SPARE TIRES

Always carry a spare tire in working order and all tools needed to change it. Check once in a while to see if it has air in it. Your wife or anyone else in the family using the car should know how to change a tire. The cost of calling a wrecker or service station truck out to change one flat tire can wipe out all of the "deal" you got on that last set of tires.

There are now some good aerosol "flat-fixer" cans on the market. When used as directed, a flat tire can most likely be inflated in mere minutes—without jacking the car up or removing the tire. Unless you are hooked on changing tires beside the road, I highly recommend that you carry a couple of these units with you all the time. The cans are not designed to really "fix" the flat—they just temporarily seal the puncture and inflate the tire. But this will get you to the next service station.

BASIC RULES FOR TIRE ECONOMY

Next to gasoline, the cost of tires is probably second in the parade of money out of your jeans. Because you don't shell out bucks every week on tires—as you do on gasoline—it's easy to forget about those four round drain-holes in your budget.

1. Buy the tires you need. Don't buy a set of expensive radials if you plan to get rid of the car next spring. When selling or trading a used car, bald tires are a penalty to the seller. You should show some tread but it doesn't matter what kind of tire it is. If you plan to run the car another 80,000 miles, then the best radials are probably the best buy in terms of long-range tire cost—provided you get them when they are on sale.

2. Buy tires when they are on sale and take the time to do some price comparisons.

3. Keep your tires properly inflated and aligned. Running at the upper limit of the *recommended* inflation range will improve gas mileage slightly and probably give slightly longer life to the tires. Don't over-inflate or underinflate.

4. Periodic rotation will almost always

Pressure cans of tire inflator and sealer are a worthwhile convenience. They can get you home or to a service station for a permanent tire repair.

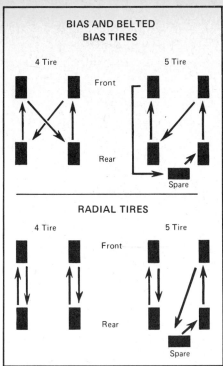

Tire rotation plans are different for bias-type and radial-type tires.

extend the life of a set of tires because they tend to wear a little differently on each wheel. Rotating the tires to different wheels tends to even out the unbalanced wear. Some tire stores offer free tire rotation service to their customers. This really doesn't cost them very much because few people take advantage of it. Take advantage of it!

5. The major cause of tread wear on a properly aligned and inflated tire is going around curves. On a tire testing track or race course where turns are negotiated at high speed, the tread can be completely worn off a tire in about 100 miles of driving. That's the same tire that can go 40,000 miles when driven conservatively.

Tire test data shows that going about 10% faster around a turn increases tread wear by about 50%.

The second major cause of tread wear is high speed. If you are driving for economy, keeping your speed down saves money two ways.

Don't speed.

Don't go fast around turns.

6. Have your new tires spin-balanced after a few hundred miles of driving. If not balanced, they hop up and down as you drive and tend to wear unevenly.

Sometimes a good balance job will last the life of the tire but sometimes tread wear changes the balance. If you feel vibration that is worse at certain road speeds, have the wheel balance checked again.

BATTERY

Like tires, buy the battery you need. If you are selling your car within the year, why buy an expensive 36-month guarantee battery? If you plan to keep the car a long time you'll probably save some money in the long run by buying a very good—and expensive—battery.

Some manufacturers are now selling batteries dry. This means the battery acid has not been added yet—this is given to you in a plastic container. This is a neat idea because if you can find a battery like this on sale, you can buy it and hang onto it until you need it. Just don't add the acid until you are ready to install the battery. If you are not competent in handling chemicals, get the auto-parts people to do this for you.

Very good service can be obtained from a battery if you don't punish it. Don't let it go dry, don't overfill it and when you do fill a battery use only distilled water. Buy a gallon of the stuff at the local grocery store and stack it with all the rest of your automotive paraphernalia.

You are also punishing your battery if the engine is out of tune and difficult to start. Prolonged cranking—especially in cold temperatures—really weakens a battery.

OIL

You can save a lot by changing the oil in your car or truck instead of letting the local garage or service station do it. Buy oil by the case—even at most service stations—and you'll get it for about half of what it normally costs when you have somebody else pour it in. Some stores often sell oil at low prices to get people into the stores. You come in to take advantage of the low price on oil and hang around to buy something they make a substantial profit on. There's certainly nothing wrong with this line of thinking. It takes two to tango.

Some auto parts stores, farm stores and truck service centers sell oil in five-gallon containers at substantial savings. Price shop. If you get serious about it—five phone calls is a pretty good start!—You'll be surprised how much the price of oil can vary.

Changing your own oil gives you the advantage of looking over your own car with an eye towards preventive maintenance. This can save you as much money as just changing the oil yourself.

If you just don't mess with it or you are living in an area or building complex that prevents you from doing these things you pay more money. If you intend to keep the car for over 50,000 miles take good care of it yourself or use a trusted garage or station at regular intervals. The long-termers need "love and care" if they are not to cost you money in their old age. The intended short-termers generally get by if you keep the oil and water to the proper level.

Changing your own oil is discussed at the back of this chapter. But let's spend a moment to discuss raising the car and precautions you should take before crawling under it. Do not short cut safety—you may join the many who are killed or injured yearly because they didn't know the dangers or were too shiftless to exercise a few steps toward safety.

Buying oil by the can or case when it's on sale can save. If you set up properly to do it, changing your own oil is not difficult.

If you are going to get under a car, jack it up and then block it up. No jack alone can be trusted. You can buy commercial jack stands or improvise as shown here. If you use cement blocks, place them with the holes pointing up.

Getting under a raised vehicle—regardless of the jack type—is dangerous if you have not blocked the vehicle so it cannot hit you if the jack fails. Bumper jacks and scissor jacks are really shaky. Do not get misled into believing a heavy shop jack is 100% safe. A shaft or bolt can shear and drop a vehicle in a split second.

If you don't have a place to dump the dirty oil just start saving the plastic containers you are currently throwing away—such as detergent, milk, shampoo and bleach. Pour the oil in these containers, screw on the cap and deposit in the nearest trash barrel. Don't lose your head and put a container of dirty oil in a trash compactor!

Everyone who knows anything at all about cars realizes oil is the life blood of the engine. Also, everyone has a little different view on the oil to use and when to change it. First and foremost, make it a practice to keep the level in the engine close to the full mark on the dipstick. Absolutely nothing is gained by running the engine "down a quart" or by adding "when the light comes on." Sooner or later you are going to pay for that quart of oil and the gamble you take by doing it later could be the price of rebuilding the engine. Labor, parts and machine work are very high-priced in comparison to bringing the oil up to the full mark.

Do not overfill a crankcase. This can raise the oil level so the crankshaft counterweights are splashing in the oil. This can throw oil up into the breather and out of the engine or into the intake manifold via the crankcase venting system. Thrashed oil has a tendency to foam which is undesirable and if you want to get picky about economy: It takes energy to splash oil. So keep it full, but don't overfill.

When to change engine oil is not quite so simply stated. Like most other phases of vehicle maintenance, changing oil for maximum economy is based on how long the car will be kept and how it is used. Perhaps this can best be illustrated by talking briefly about your neighbor Mr. Jones.

Jones is a salesman out on the road four days a week. He averages about 25,000 miles of driving per year in his leased car. Jones drives his car 24 months, then he has the pleasure of shopping for a new one.

In his appointment book he has marked when to change oil—four times a year. He has to make a quarterly report to his company four times a year and he just picks those dates for his oil changes. On the road, he always asks the service station attendant to check the oil level and he very carefully keeps the oil level at the full mark. Every other time the oil is changed, the filter is changed along with it.

Jones is pretty sharp. He knows that before clearances get sloppy in the engine and before sludge really starts building up in the engine, he will be driving a new car. Jones also knows with his long-haul driving during the week, the oil heats up so harmful sludge doesn't have a chance to form in the crankcase. Although he might get some argument from the oil companies, his economic reasoning about oil changes is pretty sound.

Jones' wife has her own car and it gets a different standard of care. A two-year old, medium-priced sedan, purchased new with the idea of keeping it for more than 100,000 miles—and maybe longer if it was still trouble-free. Properly cared for and driven, this can be a realistic goal with many makes of cars today. Mary Jones shuttles the kids back and forth to school, music lessons, visits with friends and does the scores of errands facing most housewives. Even though there is a lot of stop and go, the car only rolls up about 10,000 miles a year.

If Jones seems a little lax in caring for his leased car, he is a real bear on taking care of his family car. From October through February, Jones changes oil AND filter every thirty days. From March through September, he changes oil every 60 days and the filter every other time the oil is changed.

Jones reasons that with her driving habits, the oil usually never gets up to full operating temperature and water vapors won't boil out of the oil. Sludge and varnish can form quickly. This is especially true in the damp winter months. Jones firmly believes if you take care of something mechanical now, it will take care of you later. He plans to have the car "later" and is planning for the future.

Jones buys his oil by the case or wherever he can find it on sale. The same goes for the filters. He changes the oil himself on the family sedan—he enjoys doing it and it is easy to do while he is puttering around the garage on the weekend. It costs him about $30.00 per 12 months of driving for oil and filters.

Jones doesn't change the oil on the car he leases. He lets the local service station take care of it while he walks a block to the office and dictates the quarterly report. Jones knows maintenance and gasoline used on the leased car is a business expense deductable on his income tax. He also is not very excited about changing oil—or doing anything himself to a car he doesn't own. Last year Jones spent $32 to change the oil and oil filter on the leased car.

Both maintenance plans are sound because they fit the cars and the way they are used. If you are still in doubt as to how often to change oil, follow the recommendations in the owner's manual that came with the car. Just keep in mind that the guy writing the manual doesn't know where you live, where or how you drive. He can only outline broad rules to cover all drivers and cars.

You'll find a lot of letters on the top of an oil can these days; they have to do with the grading system of the lubricant. Prior to 1968, the best oil you could buy was marked MS. At the time, horsepower and compression ratios were going up and a better oil was needed, so a higher quality was formulated and although it still carried the term MS, the can also stated "exceeds all car manufacturers warranty requirements" or words to that effect. You may still find that on oil cans in addition to the newer nomenclature.

Under the new system of grading the highest quality of oil is marked SE. If you buy oil in bulk, there is very little difference in the cost of buying the best oil and buying something else.

Consumer data is printed on the top of oil containers but you need information to break the code. See text for more info.

Engine oil today is vastly improved over that of the '50's or even the '60's. It seems to me you should take advantage of improved technology: All sorts of additives are built into the oil such as anti-foaming, anti-wear, oxidation inhibitors, detergents and rust and corrosion inhibitors. Generally speaking, the lower the quality grade or cost of the oil, the less of the above additives you'll find in it. Whether you intend to run your bus 50,000 or 100,000 miles should have a bearing on the quality of oil you use.

The term *viscosity* refers to oil "thickness." Viscosity actually measures its resistance to flow. A high-viscosity oil is thick and a low viscosity oil is thin. Regardless of crude stock or processing procedures, all mineral oils are affected by a change in temperature. When the temperature goes down, the oil gets thicker; when the temperature rises, the oil thins out.

Viscosity descriptions are pretty simple. On the can you will see **SAE 10**, **SAE 20**, **SAE 30**, **SAE 40** or **SAE 50**. **SAE 10** is thin oil and **SAE 50** is thick oil. Although the grading system is straightforward, selecting which oil to use is not so simple. While oil viscosity changes with temperature, all oils do not change viscosity the same amount. If the engine requires an **SAE 30**—also called 30 Weight—and the oil you use thins out to **SAE 10** when it heats up, the engine could be damaged. To reduce the viscosity change due to a change in temperature, manufacturers use additives called viscosity index improvers. These additives resist oil thinning when the temperature goes up and resist oil thickening when it goes down. Thus we have another term—*viscosity index*. Viscosity index describes the *rate* at which viscosity changes with temperature.

When you buy oil, you will be faced with the choice of whether to buy a straight viscosity grade oil (SAE 10, 20, 30, 40 or 50) or a multi-grade oil. A multi-grade oil makes use of viscosity-index improvers to reduce the change in viscosity. The cans are marked 10W-40 or 20W-40 or even 30W-50. This means that at low temperature the oil will behave like a **SAE 10**, then as the oil heats up it behaves like a **SAE 40**. In other words, you get the ease of starting with a thin oil—very important when it's cold—and the protection of a thicker oil when the engine heats up to highway speed. Most engine-oil manufacturers today recommend **10W-40** multi-grade oil for the best all-around performance. This is especially true in those areas of the country where the temperature gets below 40° at night or in areas where the temperature bumps the 100° mark.

In summary, you should choose a viscosity rating suitable for the temperatures where you live and choose oil quality and a replacement schedule according to how the car is driven. If the car gets used so its engine warms up and stays there for 30 minutes or more on most jaunts, less frequent oil changes are O.K. If it is used for short hops where the engine seldom runs for more than 15 or 20 minutes at a time, change the oil monthly during the cold season and every other month when it's warm-weather time. If you do it wisely you will have the right balance between maintenance costs and service life of the machine. That's economy.

OIL FILTERS

Before oil filters became standard equipment, engines were plagued with rapid wear. It is very difficult to keep dirt out of an engine and oil and dirt whirled merrily around grinds away at bearings, crank journals and other parts. Oil filters haven't cured the problem but they have helped a lot. A filter stops small particles and holds them. When you change the filter you throw away the dirt and particles trapped in the filter.

Filters are usually easy to change and you can do it while waiting for the oil to drain out of the engine. Most cars have a spin-off filter. It looks like a tin can with a rounded bottom edge. The filter is screwed onto the bottom or side of the engine. To change it, just unscrew it, throw it in the trash and screw another one on.

You can buy filters at the local service station and auto parts stores. Like many other items you need to service a car, oil filters are often used as "leaders" to get you into the store. Substantial savings can be had by buying oil filters on sale.

Older cars have a filter element which fits down inside a canister. The canister is removed by unscrewing one long bolt which extends through the canister and into the engine. The filter element is thrown away, the inside of the canister is cleaned out with a rag and some solvent, the new element dropped in place and the canister bolted back to the engine. This is more messy than the spin-off filter but there is one big benefit. The filter element costs about one third of what a spin-off filter costs. Some engines that have been in production for many years—such as the small block Chevy V-8—started out using the element-type filter and later went to the spin-off type filter. Knowing this can save you some money. The man behind the parts counter at your car dealership might be able to find the one or two parts necessary to convert back to the element-type filter. This is a good plan if you don't mind a little extra mess and plan on keeping the car for a long time.

There are as many arguments about when to change oil filters as there are on when to change oil. One school of thought is to change filters every time you change oil. This line of reasoning is that a filter holds roughly a quart of oil and if you do not change it you are starting out on an oil change with a quart of dirty oil still in the engine. The other school of thought is that the one quart of dirty oil is of little consequence and the filter can still function properly through several oil changes.

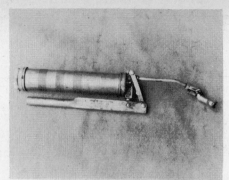

You can buy a handy grease gun like this at auto supply stores for about the cost of a commercial grease job.

The way the car is used and how long you plan to keep it should be considered for the best economic guide to changing oil filters. If you plan to keep the car for a very long period of time, you might consider changing the filter at each oil change in the winter when the filter is bound to trap a lot of crankcase condensation and then every other oil change throughout the rest of the year. If you plan to keep the car for one year and trade it off, changing the filter twice will get you to the end of the year.

Use your new grease gun one time and you have paid for it.

Like oil changes and grease jobs, oil filters will not give you better gas mileage. All of the items mentioned however will lead to less wear and tear of parts which can plague you with repair bills if the car is kept for an extended period of time.

GREASE JOBS

One of the economic beauties of doing your own routine maintenance on your car is knowing more about it than anyone else. Consider the lowly grease job. Over the years the number of lube points have gradually been reduced. On most cars built now, only the ball joints and tie-rod ends need lubing. By lubing the ball joints on a regular basis you not only gain peace of mind that the job was done and done correctly, but you know what to tell the mechanic when he tells you the ball joints need to be replaced—after 25,000 miles. Properly lubed ball joints and tie-rod ends will last about 100,000 miles on a modern car. A hand-held grease gun and cartridge of grease and one Saturday morning every month or so is very low-cost insurance for trouble-free driving and to keep from being sold some front-end parts you didn't need in the first place. Even the poorly written owner's manual that came with the car will tell you where your car needs to be lubed.

AIR CLEANERS

If you can be lax—even negligent—about oil filters, oil and grease jobs and not see gas mileage suffer then the same should be the case with air cleaners, right? Hardly! Wear of the engine increases greatly if the filter is missing or if it is

This oil filter was opened up and the filter element cut in two to show the amount of sludge and abrasive grit that collects in there. When filters plug up, they no longer function and the grit is not removed from engine oil.

simply not filtering. If the filter is badly clogged, gas mileage can suffer.

There are three kinds of filter elements. Historically, the oil-bath filter comes first. This is no longer used on passenger cars.

Most filters are a throw-away paper element or a washable-foam element. There are variations. There is a throw-away paper type. There is a washable-foam slipped over a paper element. This is the most efficient filter you can buy, but it costs a lot.

One recommended service procedure for a throw-away paper element is to replace it every 10,000 to 12,000 miles. Several times during its service life you are supposed to bang it on the ground to shake out some dust and dirt or blow out the element with an air hose. I can't get excited about either procedure. Keep in mind you are dealing with paper which is a fragile material.

Even manufacturers are now backing off the air-hose procedure. Look closely at the surface of a paper element and you'll note that the paper is fuzzy in texture. Blasts of air can fold these tiny fibers over the pores in the paper—thus clogging the filter that you wanted to clean.

My suggestion for paper air-cleaner maintenance is change the element more often than the manufacturer recommends. If much of your driving is off the pavement—or you live in an area with a lot of dirt in the air due to agriculture, road construction and such, change filter elements twice as often as the factory recommends.

Within the past several years, open-pore polyurethane foam has become a respected filter material. The beauty of this material is that when saturated with a light oil it is an efficient filter and it can be cleaned and reused over and over.

When replacing either foam or paper or a combination of those elements, it is a good idea to smear a light coat of wheel-bearing grease on the upper and lower seating area of the element and also make certain that the element is perfectly seated before finishing the job. The light film of grease will effectively seal out any air and dirt that might try to slip between the filter and the filter housing or cover.

Air-filter elements can be purchased from auto parts stores. Filters are often on sale and substantial savings can be

Most air filters open up for servicing by unscrewing a wing nut on top as shown here. This "hot rod" filter uses a chromed wing nut drilled with holes for lightness in the true racing tradition. Although fine for show, this type of small air cleaner doesn't do as good a job as the stock cleaner on your car.

Old-style oil-bath air cleaners have a reservoir in the bottom (arrow) which should be kept full of engine oil. Better to replace this antique with a modern paper-element cleaner.

Some service manuals recommend banging a paper-element filter and blowing it out with an air hose to remove accumulated dust. What can happen is shown in the next photo.

After banging and blowing, you may wind up with this. That wire mesh will stop owls and tree branches but not much else.

Greasing the edges of a new filter element before installing helps persuade all the air to go through the filter.

made by buying while they are on sale.

One kind of filter should be avoided regardless of cost. Small, "hot-rod type" filters that show off the engine simply do not have sufficient area to effectively clean all of the air consumed by the engine. The air travels through these small filters at such high velocity that dirt literally hurtles straight through the element—right into the engine. Stock, large filters offer more filter area. Stick with this size filter, it is your best bet for maximum gas mileage and less engine wear.

SPARK PLUG REPLACEMENT

A lot of money can be wasted on—and because of—spark plugs. A set of plugs represents far more of an economic stake in a car than just the purchase price of a new set.

Before you ask—there are no miracle, lifetime, double-gas-mileage, triple-horsepower, super-trick spark plugs. Save your money and buy reputable brand plugs. A spark plug won't last forever, but they will last far longer than some garages and parts salesmen would have you believe.

The place to buy plugs is wherever they are on sale. Stores often have them on sale and you can save 20 to 30 percent. Don't buy used reconditioned plugs. The chance of very short life or a "dead" plug is just too great.

You don't necessarily save money by changing plugs at any particular mileage interval such as 10,000 miles. A set of plugs in good working order after 20,000 miles is not uncommon.

In some research several years ago, AC Spark Plugs showed that properly functioning plugs can mean a great deal to mileage. On a V-8 engine, AC found that if one plug is misfiring half the time mileage can fall off as much as 5.5% at 30 MPH. As speed rises, the loss becomes greater. At 60 MPH, the mileage can drop off by 7.3%.

This means if all the plugs are firing all the time and you are getting 18.2 MPG at 60 MPH, one misfiring plug can cause a drop down to 16.7. If two of the plugs are misfiring half the time, gas mileage can drop off as much as 16% at 30 MPH and 19% at 60 MPH. Now you are down to 14.6 MPG, all because of two bad plugs. With three plugs misfiring half the time, gas mileage at 60 MPH can drop 30.7%. That means your clunker is now getting 12.5 miles-per-gallon instead of 18.

The point is, you can buy a new set of plugs with the money you will save on gas once the new plugs are installed and have money left over.

If a plug is not firing or is misfiring, remove it and clean it. Many service stations and garages have small grit-blasting machines designed just for spark plug cleaning or you can clean your own by soaking them for an hour or so in benzine—sold at most drug stores—and then carefully cleaning the crud off the insulator and electrode with a small, soft wire brush. Then rinse the plug in clean benzine and allow it to dry before regapping. If the plug continues to be dead or misfire, replace it with a new one.

When regapping spark plugs, gap a little on the wide side of the recommended gap for a little better gas mileage. In other words, if the engine specs call for a gap of 0.035-inch, open the gap up to 0.038 or even 0.040-inch. If the plugs misfire, bring the gap back down to spec—but if the plugs fire, you'll benefit from a slight increase in gas mileage. You will have to regap more often though.

Keep in mind that if you change plugs before they need it, the cost of plugs can add up in a hurry. On the other hand if your engine is carrying around a couple of plugs that just fire now and again, good gas mileage is going right out the tailpipe.

BETTER GAS MILEAGE WITH SUPER ZAPPY IGNITION?

If you are raising the question about getting better gas mileage by ordering one of the many aftermarket ignition systems now on the market, at least hold on to your money until you have read our full test report on one of the units in the next chapter—*Saving is Not Spending Anything*.

ANTI-FREEZE

Modern anti-freeze does a lot more than keep the coolant from freezing in the winter. It makes the engine run a bit cooler in the summer, is not as prone to boil as water, doesn't evaporate as quickly as water, prevents corrosion, lubricates the water pump, and some of the stuff will even seal up tiny leaks in an old radiator or prevent seepage around freeze plugs in a block. Buy it on sale, keep a gallon or so on hand and when the coolant level is low, add some instead of water. With emission-control equipment and modifications making engines run very hot, anti-freeze in the summertime is essential to the life of an engine.

BRAKE FLUID

Like anti-freeze, try to find brake fluid on sale and keep some on hand at all times. The average car owner won't use a gallon of it in a lifetime. When you are doing your Saturday morning checking under the hood, pop the top off the master cylinder and add fluid if it is needed. When bleeding brakes, catch the fluid coming out of the wheel cylinder to keep it from making a mess in the garage or driveway, but don't pour it back into your supply of new fluid. There is a difference between being frugal and unwise.

TRANSMISSION FLUID

Before getting carried away and buying a case of automatic-transmission fluid on sale, find out what kind you need. The owner's manual, a service manual for your car or the transmission dip stick will tell you what type to use. This is the only kind that should be used in that particular transmission. Using the wrong kind can shorten the life of the innards greatly.

Check the fluid level exactly like the owner's or service manual says—normally this is with the engine warmed-up and running, the transmission in Park and the car sitting level. All this must be followed to the letter or you'll get a false reading.

If the level is quite low, don't immediately assume it will take an entire quart and dump one in. Add no more than a cup of fluid, and check the level again. Repeat the procedure until the level is brought up to the full mark. Never overfill an automatic—most will blow out the excess—and then some—when they get hot and have been overfilled. This is messy, smells like the car is on fire and gets you right back where you started—with a low fluid level.

WINDSHIELD WASHER FLUID

If you live where it doesn't freeze, add a little household window cleaner to the window-washer bottle, then fill it with water. The mixture will give you cleaner windows than straight water and does a better job of getting insects off the glass than water. If you live where it gets way below freezing, you'll have to mix a pretty strong solution of window cleaner and water—about 50-50—to prevent it from freezing. Depending on the price, you may find it more economical to buy windshield-washer fluid.

How to be a fair-to-middlin' gas-pump attendant

This is the special section for those who have been denied the educational benefits of poverty—until now. If you have never known the quiet satisfaction of topping up the brake fluid, or the envious glances from your neighbor while you are draining motor oil on the sidewalk, here is your chance.

The real confidence-builder is to get the hood open some way. You have to find two latches. This car has latch No. 1 inside the driver's compartment.

On other cars, the first hood latch is in the grille or above the license plate as shown here.

When you find the first hood-release control and pull it, the hood won't say "AH" but it will open up about an inch. Look or feel in the crack and you will locate the second hood release latch—a safety catch.

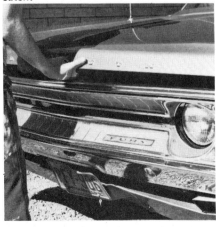

With the hood open, you can get in there and look around. Identify the parts you intend to service and give the underhood area a good looking-over for things that aren't right.

As you can see in this photo, all parts are easy to identify and there is plenty of room to work. That's a tiny joke. This is a Rolls Royce. If it happens to be your car, stop reading this book!

Remove the radiator cap and check fluid level by looking. Coolant level should be just below the bottom of the filler neck. Modern radiator caps have a click-stop position halfway between fully on and fully off. When you feel the click-stop, leave the cap in that position while radiator pressure escapes. If steam comes out, be extra careful. It's best not to try to service a hot engine. Do something else for a while.
If coolant level is low, fill to bottom of filler neck with antifreeze.

Even radiator caps aren't simple any more. They have an internal valve which should open up at a specified pressure and allow steam to escape. Pressure rating is marked on cap—in this case 12-15 pounds.

Pressure caps keep coolant from boiling. If your car boils, have the radiator cap pressure-checked at local service station or garage. Pressure tester looks like this. If it doesn't hold rated pressure, replace cap. It's a good idea to have the cap checked once a year even if it seems O.K.

Remove battery filler caps and check fluid level by looking—it should be just below filler openings. Battery terminals here are nice and clean. In tune-up chapter, you'll see what to do about corroded terminals.

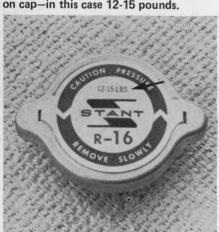
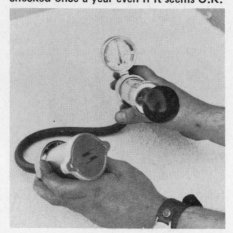
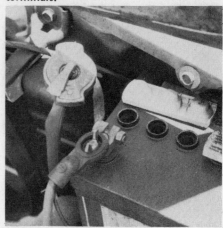

 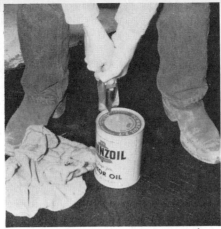 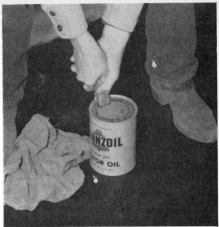

If battery fluid is low, top up with distilled water. If you have nerves of steel, you can pour directly from distilled water container and get some in the battery. Otherwise, use kitchen measuring cup with pouring lip or a plastic funnel. Battery has acid inside which will get on tip of funnel. Don't get any on your clothing or it will make a hole. Rinse tip of funnel after use. Never add acid—only distilled water to make up losses due to evaporation. Don't smoke while peering into battery openings. There may be an explosive gas in there and if so you get a face full of acid.

Buy a professional oil can opener and pouring spout at auto parts stores. Force the gadget into the top of the oil can.

Force the pouring spout into the can as far as it will go, so you get a good tight seal between can and spout. It will leak anyway. Cowboy boots add a little class to the operation but are not essential.

Pour in some oil, check dipstick again. Don't overfill. Some dipsticks are marked so it takes one quart between the ADD OIL mark and the full mark, but don't count on it. Add a little and check, until you get the feel of it. Use a rag to catch spillage and wipe off the engine when you are through. Spilled oil on a hot engine makes smoke, bad smell, messy working conditions.

On top of the engine is a oil filler cap. If it isn't obvious, watch where the service station person pours oil. Then you'll know. Remove it so you can add oil.

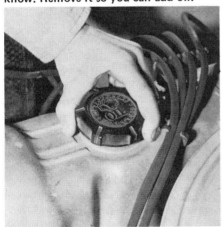 

73

Here's your basic expert's kit for changing engine oil. Wrench to remove oil drain plug on bottom of engine, pan to catch the oil as it drains out, old plastic bottles to receive drained-out oil so you can donate it to the trash collector without him knowing it, top of plastic bottle cut off to make cheapo funnel, oil can spout and a precision wiping rag.

Wiggle under the car or block it up safely as shown earlier. Remove drain plug with catch-basin at the ready position. If drain plug falls down into pan, fish it out later. If there is a gasket on drain plug, don't lose it. Let all the oil drain out, *replace drain plug*. Then pour correct amount of oil into filler on top of engine.

Large opening in oil-filter wrench fits over body of oil filter and grasps it tightly so you can unscrew the filter cartridge. Moisten rubber gasket on new cartridge with engine oil and screw it in place. When you install a new filter it takes about one quart more oil to fill the engine because you also have to fill up the filter cartridge.

Unless you own a gorilla, you need a tool that looks like this to remove a spin-off oil filter cartridge. At the auto parts store, ask for an oil-filter wrench.

It's bad news to need your windshield washer and find it out of fluid. Check occasionally and refill. The tank will have a hose line leading back toward the dash or windshield.

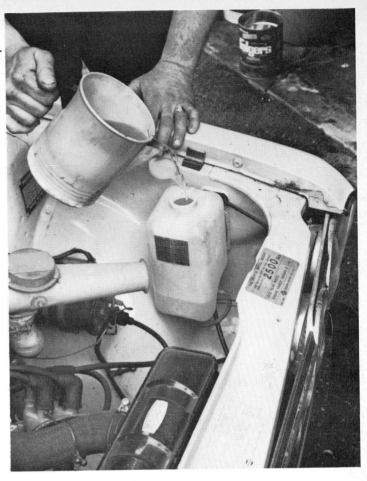

Some radiators have a sealed-on cap and an overflow tank which receives overflow coolant when the engine is hot and feeds it back into the radiator when it cools off. It has a hose leading over to the top of the radiator. Filling is done by bringing the fluid level in the overflow tank up to a NORMAL mark on the side of the tank.

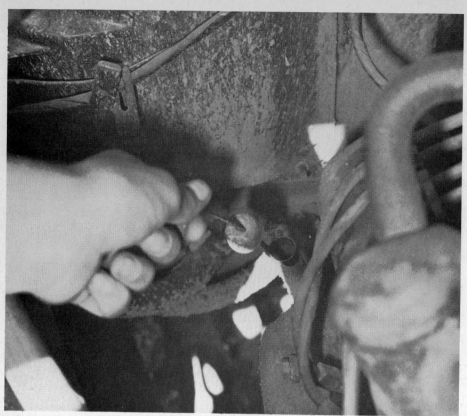

Cars with automatic transmission have a long dipstick located back near the transmission. You read it the same way you do an engine oil dipstick. Be sure you are following correct procedure such as running the engine at idle with the transmission in neutral or whatever is appropriate for your car. Check the owner's manual.

If you need to add transmission fluid, be sure you add the correct type and do not overfill. A special funnel with a long neck helps you do an otherwise impossible job. Funnels are available at auto parts stores.

5 RIP-OFFS
Saving is not spending anything!

An extreme fire hazard exists on cars equipped with too many economy improving devices. If the total of all the economy improvements added to your car exceeds 100% there is a danger of the engine *actually* saving more gas than it uses. This gas could back up in the gas tank and finally spill out—causing a *great fire hazard!* This is not quite so important on cars equipped with devices that improve performance along with economy, because these cars can easily accelerate to 300 MPH in 3 seconds and outrun the fire.

If you believe the above paragraph you will believe every one of the claims for trick mileage-improving devices.

It was P. T. Barnum of the old Barnum & Bailey circus who said something like "There's one born every minute," and became a folk hero by flagrantly suckering the population and admitting it at every opportunity.

"People like to be fooled," they say. You can decide about that for yourself, but when the foolery transfers bucks from your wallet to someone else's, it usually is more fun for the other guy.

Popular magazines are filled with ads for new miracle gadgets that promise gas-mileage increases of up to 50 percent or more. There are only two words in those promises that mean anything—the words "up to" which you should instantly translate to mean "less than!" An economy improvement of exactly zero is "up to" any percentage you want to name.

As you will see in this chapter, with my associates at Doug Roe Engineering we made actual road tests on some of these modern miracles and found that some of them produce gas-mileage increases of "up to" zero.

SPARKLE PLENTY— BUT DOES IT IMPROVE ECONOMY?

The automotive ignition system has been the target of many "economy improving" devices over the years. Among the large number of these devices many of them do the same thing and can be grouped accordingly.

SPARK INTENSIFIERS

These items theoretically raise the output voltage of the ignition coil. The theory is the hotter spark from the coil will make the cylinders fire better, increase fuel economy, decrease emissions output, increase power and clean the combustion chambers. The spark intensifier (sold under a variety of names) is usually nothing more than another gap in series with the path from the high voltage side of the coil to the spark plug. How many of these claims are reasonable? Well, let's see the test results.

SPARK INTENSIFIER TEST RESULTS

The spark intensifiers cost $1.99 for a set of four from a large mail-order firm. Installation was easy—the small intensifiers were plugged on top of the spark plugs and the ignition wire was in turn plugged into each.

The intensifier is two small pieces of metal separated by a plastic section. They

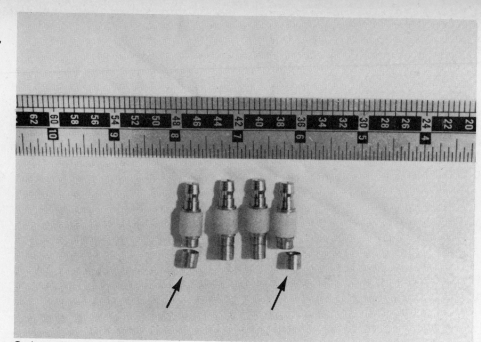

Gadgets like this are sold under a variety of names and promises as a way to increase fuel economy. These installed O.K. to allow testing, but two of them broke when being removed from the spark plugs. See the curve for our test results.

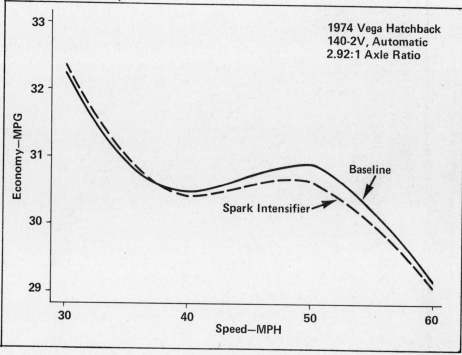

Testing showed a slight improvement in miles-per-gallon at low speeds and a slight decrease at higher speeds, however, the changes are considered to within the margin of error in normal test procedures. Conclusion: Spark intensifiers have no measureable effect on fuel economy.

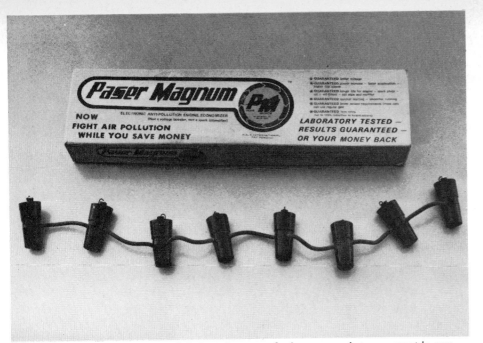

This "mixture ionizing" device failed to show any fuel economy improvement in our tests. The set illustrated above was also evaluated at a sophisticated computer-monitored emission facility. It had no significant effect toward reducing CO, HC or NO_x in these tests.

Replacement ignition distributors and electronic ignitions of several types are available as bolt-on accessories. No single evaluation can be given for all brands and types because quality and applications vary. See the text for more information.

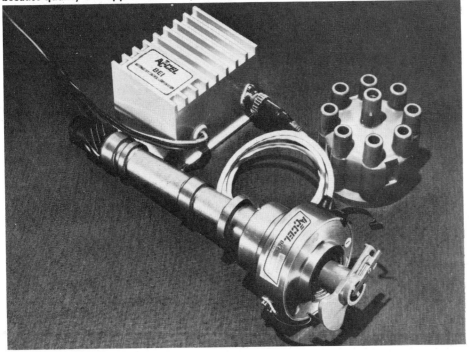

were installed on one of our Vega test cars with an automatic transmission. We established baseline mileage figures before installing the intensifiers and then made tests. Any change caused by the device falls within the 2% error limits of our testing. In the case of the intensifier the difference in MPG amounted to less than 1%—roughly 0.3 MPG.

Acceleration tests conducted with the intensifiers on the same vehicle showed even less change. After conducting more than 36 separate runs to verify our testing we conclude the spark plug intensifiers have no effect on fuel economy. Our acceleration tests measured no difference in fuel economy and only 0.05 second difference in time for a 0 to 60 MPH acceleration run!

With the amount of money we spent on the intensifiers we could have purchased four gallons of gas at the time of the test—a far better buy.

MIXTURE IONIZER

Another type of ignition system add-on device is the so-called mixture ionizer. Manufacturers claim much the same benefits from these gadgets as the spark intensifier people do. Basically, the device is supposed to take part of the energy which is going to the firing spark plug and channel this energy to the other seven plugs—in an eight-cylinder engine—to "ionize" the mixtures in these other cylinders. The "ionized mixture is then supposed to burn better, more completely and generate more power than if the "ionization" process had not taken place. Our information and test data indicates none of these claims are true under normal driving conditions.

CAPACITIVE-DISCHARGE IGNITION

Another type of add-on ignition device is the capacitive-discharge (CD) ignition. These devices have some economic advantages over the conventional ignition system, although fuel economy is not a major one. Accessory electronic ignitions use the ignition breaker-points to trigger the electronic circuitry which supplies the current to the coil.

There are two advantages to this set-up: The points no longer carry the coil current so burned points are a thing of the past and the electronic circuitry can provide more current to the coil so the output voltage can be significantly higher than the conventional system.

What this means from an economy point of view is that tune-ups will be less frequent and less expensive when they are needed. The points will last as long as the rubbing block stays lubed and does not wear out. All the electronic systems need as a trigger is for the points to open and close, so point-dwell adjustments are no longer needed. These systems do not use an external condenser so this item and the time needed to install it can be removed from the tune-up bill. Spark plugs last longer because the higher available sparkplug voltage will fire fouled plugs the standard system won't fire.

This brief description may indicate that an electronic ignition system is the panacea for all your car's ignition ills, but there are several other points to consider. The electronic systems usually put out a single pulse of high energy, but for a very short duration. This can cause misfire in an engine set up with lean mixtures during idle and part throttle for emissions control. In fact most of our exhaust emission tests on engines with aftermarket CD ignition systems have shown poor results. Lately several manufacturers have developed multiple-firing CD systems which hopefully improve the emission performance of CD systems.

Another of the considerations before buying an add-on electronic system is the possibility of poor workmanship or faulty components which lead to failures in the system. There are a certain number of manufacturers who have cut quality standards to the point it seems very unlikely that their products will last 10,000 miles, much less the life of the car. Before buying, get the opinions of others who own the brand to find out if it is one of the good ones. One desirable feature of some of the electronic systems is provision for a switch which allows you to convert back to a standard ignition system if the electronic circuit fails. Even the best of the electronic systems can have a freak component failure, so the switch is a good thing to have.

CAPACITIVE DISCHARGE IGNITION TEST RESULTS

The particular unit tested for this section was a single-firing CD system capable of putting out about 40,000 volts at the coil. Forty thousand volts is a mighty impressive sounding number of anything—but what does it all mean?

Higher spark voltage will fire a nearly fouled plug to save you money by extending the interval between tune-ups. The points last a long time because they no longer carry the full current for the ignition system. And, theoretically, high-energy spark is supposed to increase gas mileage, power, performance and acceleration.

For test purposes we installed the accessory CD ignition—$29.95 from a large automotive mail order firm—on our 1973 Chevrolet test truck. When installing the unit we mounted a switch on the dash allowing us to select conventional ignition or the CD unit at the flip of the switch. This eliminated installation and removal times so we could make all the baseline and test runs in a short period. Weather, temperature and any number of other variables creep in when a lot of time passes between the baseline and test runs.

The accessory CD ignition was worse over the entire test range—4.7% at 60 MPH and 5.3% at 30 MPH. This is not surprising if you consider the vehicle and the standard ignition system. Late-model emission-controlled engines with lean carburetors, Exhaust Gas Recirculation and other associated hardware use mixtures in the combustion chambers that are difficult to fire and, once fired, they won't necessarily continue burning until combustion is complete.

The stock ignition system compensates for this by the natural phenomenon of "ringing" in the coil which will fire the plug as many as six times.

CD ignitions, in standard form, only fire once. This is a high voltage and very powerful spark, but if the fire doesn't stay lit—there is no second chance to relight the mix. On older model engines with richer part-throttle mixtures, this CD unit would probably equal the economy of the standard ignition.

The CD ignition equaled the standard system on the 30 to 80 MPH acceleration runs and was worse at all speeds in fuel economy. Rich, dense mixtures in the combustion chamber will keep burning after being ignited by a single spark, if the spark lasts long enough.

WHAT ABOUT OTHER CDI UNITS?

These tests on an accessory ignition do not tell the story about all CDI units. There may be better ones available for bolt-on. Because there are a lot of brands

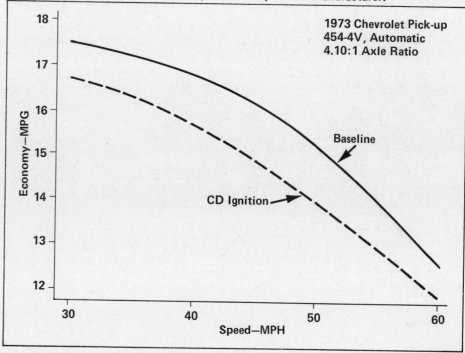

Testing of one mail-order CD ignition showed reduced fuel economy compared to the same car with stock factory ignition of the non-electronic type. These results do not necessarily apply to all brands and types of electronic ignition, but caution is advised unless the electronic ignition is factory-installed by the car manufacturer.

CAPACITOR-DISCHARGE IGNITION PERFORMANCE—30 to 80 MPH Acceleration

Test	Time (Sec.)	% Change	Fuel Consumed (Gal.)	% Change
Baseline	29.98		.0899	
CD Ignition	29.98	0	.0893	-0.67

Under hard acceleration, the CD ignition tested was equal to stock in elapsed time and slightly better in fuel economy. CD ignitions can give better results when used to fire richer mixtures such as during hard acceleration.

Building your own water-vapor unit can save you some money whether it works or not. If you try one of these, be sure to keep the container filled with water.

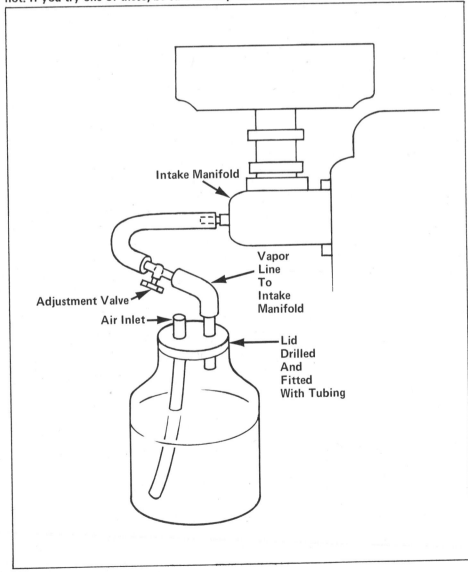

and prices, we can't rate them all. I can advise you to check carefully before laying down your bucks—preferably with somebody who has already gone the route and tried the brand.

The situation is also a little different with factory-installed CDI units. I discuss this later in the chapter on buying a new car.

WATER INJECTION

There seems to be an innate desire among people to believe a simple piece of hardware can increase gas mileage and horsepower significantly. The prospect of burning water in an internal-combustion engine must be placed at the zenith of hope and optimism. Hucksters peddling water "injection" devices use random half-truths that make for good advertising copy in the hands of a skillful writer. "You've noticed your car runs better when it's raining . . . . World War II airplanes increased horsepower and range by burning water . . ." and so forth.

Water-vapor devices can be of limited benefit to some engines in some cases—but the possibility of them helping gas mileage is dubious. First, it should be understood that the current crop of "water-burning" devices do not inject—regardless of what the brand name or promotional flak would lead you to believe. Water vapor is drawn into the intake manifold by vacuum—the higher the vacuum, the more vapor drawn into the engine. High vacuum comes at idle and when the throttle is backed-off while cruising. Elsewhere in this book you will see how little fuel is consumed during idle and closed-throttle operation. Thus, the simple hardware of vapor devices is at its best when the engine is burning the least amount of fuel.

On engines with chronic detonation (pinging) problems, even the slight amount of water vapor drawn into the combustion chamber during part-throttle operation can prove to be of some help—in suppressing detonation—not improving gas mileage.

Water-vapor devices most always promote the use of some additive at an inflated price to be added to the water. Think about it for a minute. Four ounces of anything that costs a buck had better deliver whopping gas mileage to pay for the gas you could have purchased for that dollar. The additive is always some form of alcohol or acetone with a little color-

Homemade water vapor device manufactured from a pickle jar proved to get "up to" the same improvements in everything as offered by commercial units.

This device fits between a 4-barrel carburetor and the engine. The wire "socks" hang down into the intake manifold beneath the two smaller carb barrels and do something. Whatever it does is evidently not necessary for the large carb barrels when they are opened up to gulp in air and fuel. Something gets lost—*your money*—between advertising claims and performance.

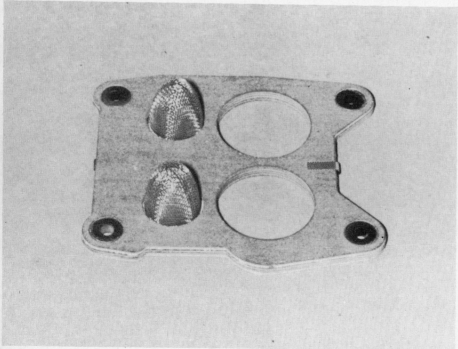

ing added for effect. Race cars, boats and airplanes have relied on alcohol as fuel for years.

But what most people don't understand is the injection or carburetion for an engine that burns alcohol must be set up to burn alcohol—which is consumed in far greater quantity than gasoline in the same engine. In other words horsepower is increased by using alcohol for fuel but the pounds of fuel being consumed per hour is much higher.

Under certain driving conditions—a lot of stop-and-go driving and long idle periods—a water-injection device may help keep the combustion chamber and spark plugs free of carbon buildup on certain engines. We could not measure ANY change in performance or gas mileage with a water-vapor device.

They are easy to construct and cheap for those who like to tinker and do their own work. If a vapor device intrigues you, build your own and use plain water in it. This is far better economy than buying one and paying money for alcohol-based additives.

BUILD YOUR OWN WATER-VAPOR THING

Building and installing your own vapor device is a low-cost project for those interested in experimenting with the possible benefits.

A fairly large clear or translucent plastic container is needed to hold the water—a windshield washer bottle obtained from a wrecking yard works fine because they usually have some built-in provision for mounting. Two holes are needed in the lid to accommodate two lengths of 3/8-inch steel or copper tubing. Details are shown in the accompanying sketch. One long piece of tubing should be epoxied into the lid with the tubing installed so it reaches nearly to the bottom of the water bottle. The other length of tubing should be quite short. To the short tubing attach rubber fuel line and some sort of adjustment valve. Surplus, hardware or plumbing shops can supply you with a small in-line valve. Route the rubber fuel line from the adjustment valve to a fitting on the intake manifold. Unless you can find or adapt an existing fitting on the manifold, you'll have to drill and tap for one.

Make sure all steel line to rubber hose junctions are tightly held with hose clamps. Mount the bottle on an inner fender panel or firewall so the water level is *below* the inlet on the intake manifold. Fill the bottle to within one inch of the bottom of the short tubing routed to the intake fitting. If the water is difficult to see in the bottle, add a drop or two of food coloring.

Turn the adjustment valve off. Start the engine and allow it to idle at normal speed. Slowly open the adjustment valve. You'll notice a stream of bubbles coming from the bottom of the long piece of tubing going to the bottom of the bottle. Adjust the control valve until a steady stream of bubbles spaced about an inch apart comes from the tubing while the engine is idling. Place a drop or two of rubber cement or gasket sealant around the adjustment-valve shaft so vibration won't alter the adjustment.

Keep track of the water level for several days until you develop a feel for how much water is being consumed. Keep in mind that if the bottle runs out of water, you'll have a substantial vacuum leak which is potentially dangerous to the engine. After installing a water-vapor device it is normal for the idle speed to increase—simply lower the speed back to normal with the idle-speed adjustment screw.

MINI-TURBOCHARGER

Advertising copy used to sell the mini-turbocharger is far more interesting than the $9.95 device. Their literature says, " 'your' engine runs best only when it idles!" The imaginative writer allows, "The modern carburetor is an idiot that hasn't had a major advance in principle in 50 years!" The other side of the copy mentions "28% more power . . . save up to 2 gallons of gas every hour . . " In short, the promotional effort surrounding the device is the classic approach of dazzle 'em with your footwork, baffle 'em with your bull - - - -.

The little aluminum unit is easy enough to install—a feature of all the cure-all mileage devices. You simply insert it in the PCV line. There it creates a vacuum leak which lets more air into the engine. Because vacuum is highest at idle, that's where the biggest leak occurs. As a result it caused our test car to idle very roughly with a vacuum-gage reading lower than in stock form. Even in stock form the Vega carburetor delivers such a lean mixture

Perhaps the most ancient of the mileage-improvers is an air leak into the intake system. They come in various types and styles with mind-boggling names such as *Mini-Turbocharger.* **At least, it's mini.**

Disassembled, the *Mini-Turbocharger* **reveals spring-loaded drilled plates which evidently bleed air into the intake manifold according to intake vacuum. The piece of foam appears to be an air filter.**

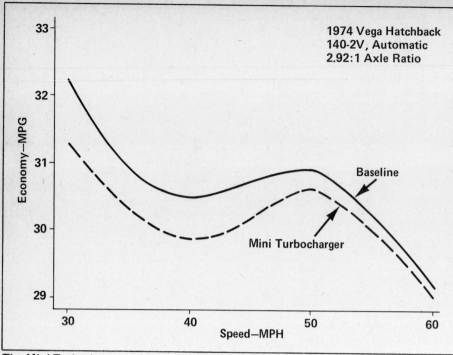

The *Mini-Turbocharger* turned in a solid decrease in fuel economy at all speeds.

that the car surges at steady 30 and 40 MPH running. The mini-turbocharger made this condition worse. Level-road economy decreased when the device was installed and was worse at slower test speeds.

At wide-open throttle during acceleration tests, the device simply does not operate. As a result, our 0 to 60 acceleration times did not change, nor did fuel consumption.

Although the ad copy makes the simplistic and highly inaccurate statement that an engine runs best only when it idles and that no improvements have been made in carburetors in the past 50 years, one size mini-turbocharger is supposed to be effective on any size engine.

Time and again the ads allude to supercharging and turbocharging, both of which are proven methods of increasing horsepower, if not mileage. This is neither a supercharger nor a turbocharger and it does not in any way use the principles of either. It is an air leak, period! The only way one of these could help economy would be if the carburetor was running too rich in the first place.

THE GR (GET RICH?) VALVE

The GR valve inserts in the PCV vacuum line between the PCV valve and the carburetor. It looks like a plastic Mini-Turbo Charger and is another form of air leak. The tests results are almost identical. We tested the GR valve because of the large amount of advertising they were doing and because it was endorsed by Gordon Cooper—the astronaut and scientist. We assumed, as many people would, that a man of the stature and scientific ability of Col. Cooper would not endorse a product unless it had real merit.

The claims for the GR valve were all in the area of economy. One of the ads claimed a 172% increase in fuel economy on a 1966 Pontiac. We used a 1974 Vega and a '69 Pontiac GTO to test the GR valve. The test data obtained from the Vega can be compared to the information we obtained from running the Vega with the Mini-Turbo Charger and the Pontiac data can be compared to the *economy claims* they made for the GR valve equipped '66 Pontiac.

The Vega tests show much the same results obtained with the Mini-Turbo Charger. Economy was decreased at all tested speeds with the GR valve and

Response to impassioned advertising got us this *GR Valve.*

showed a maximum decrease of 4.8% at 40 MPH. The test Vega with the factory 2-barrel carb was already very lean—in fact the car had objectionable lean surge at 35 to 45 MPH speeds. The GR valve further leaned the mixture and the surge became much worse.

The much larger engine of the GTO seemed to have fewer ill effects from the "vacuum leak." The idle was not affected as much as it was on the Vega and light throttle driveability suffered only slightly; however economy was down substantially at low speeds.

RIVERSIDE AIR-FLOW NEEDLES (SAME AS GANE)

The Riverside Air-Flow Needle is a replacement idle needle for the carburetor with a hole drilled down the center. The hole is intended to bring in a stream of air to mix with the fuel coming through the idle system in an effort to achieve 10 to 25% more economy, keep spark plugs cleaner and so forth. Literature included with the device makes totally inaccurate statements about how a carburetor works and why this device will improve its functions. We tested the Air-Flow Needles on a Pontiac GTO and on the new Bricklin automobile which became a part of our test fleet while information for this book was being compiled.

The Pontiac reacted very poorly to the installation of the Air-Flow Needles. The best idle we could attain was 1.5 inches lower vacuum than the previous (stock) idle with the standard needles. The idle was rough and light-throttle accelerations from a stop created a heavy surge. This surge was also evident to a lesser extent at the 30 and 40 MPH level road speeds. Economy was down at all speeds with the worst drop being at the 30-40 MPH range.

The Bricklin has a 360 CID American Motors V-8 transplanted directly out of a '74 Matador. This engine is typical of many emission controlled engines in that the idle and off-idle circuits of the carburetor are quite lean. Surge was so bad at 1400 RPM that on a gentle acceleration, the engine would reach that speed and then slow down slightly as the driver continued to press the accelerator. Finally, the carburetor would be opened far enough to pull fuel from the main metering system. At this point the vehicle would accelerate briskly.

The Riverside Air-Flow Needles made

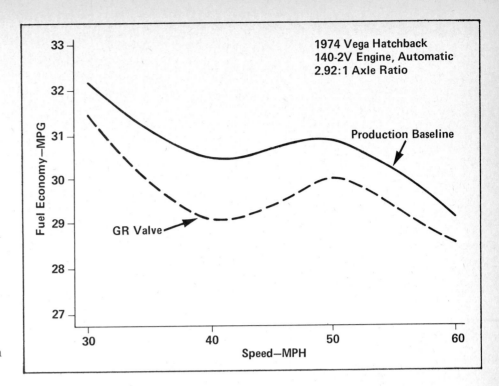

Testing the *GR Valve* got us these results.

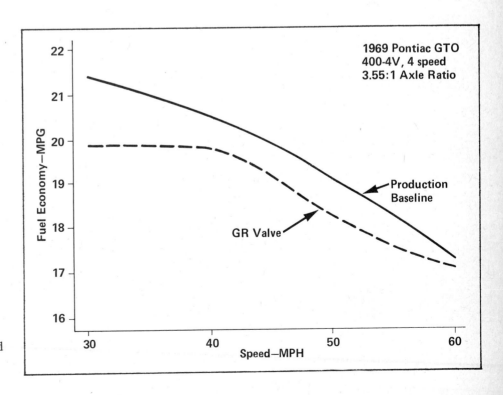

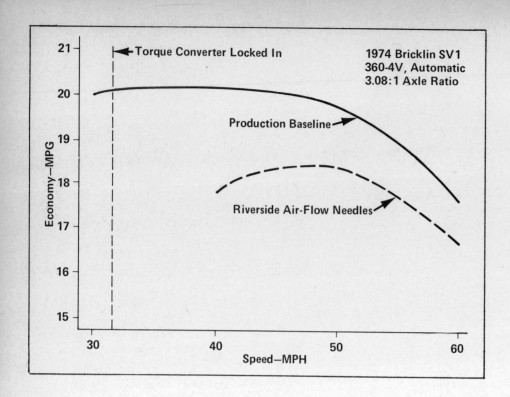

These are the test results on the modern air-leaking needle.

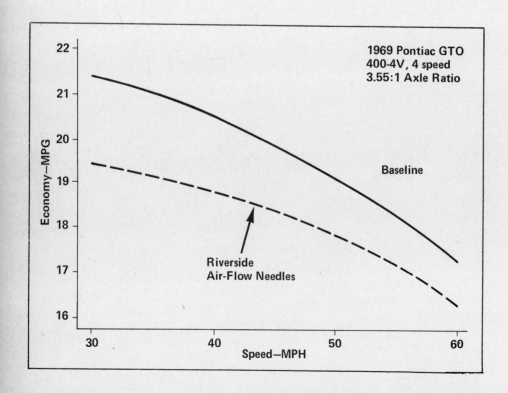

86

the Bricklin almost undriveable at certain speeds. Idle was poor and the hesitation evident in the engine in stock form became intolerable with the installation of the Air-Flow Needles. The problem was so bad that we could not run economy tests at 30 MPH because the engine refused to act civilized at 1300 RPM in high gear. Economy also decreased on this engine—especially at the lower speeds.

MINI-JECTOR

Fuel-pressure regulators are sold under a variety of names—Mini-jector and Super Energy Cell are just two examples—but the best of the regulators are simply called fuel-pressure regulators. Fuel-pressure regulators by any name are simple devices that fit in the fuel line between fuel pump and carburetor.

The unit is supposed to even out the fuel pulses created by the pump. The theory behind the sales pitch is lower and more even fuel pressure pulses let less fuel into the carburetor and prevent flooding, dying and excessive gas consumption. If you continue to believe the ads you'll get more pep . . . instant response and other benefits.

In actual practice a good pressure regulator *will* even out pulses of fuel pressure going to the carb and *can* prevent flooding and give somewhat better gas mileage if, *IF,* there is a problem to begin with. IF the float in the carb is out of adjustment or IF the needle and seat are damaged or badly worn, then a pressure regulator can be of some benefit. At the same time you should know that a pressure regulator can restrict the flow of fuel so the engine surges or stumbles at near wide open throttle operation simply because it is starving for fuel. The Mini-jector we tested was adjustable as are a number of fuel pressure regulators. Unfortunately, the Mini-jector labels its calibration A, B, C, D and E which makes it impossible to determine what the exit pressure is in terms of pounds per square inch (PSI).

A better approach is taken on a unit called Filt-O-Reg which is also adjustable but tells you what the exit pressure is in 1/2-pound increments from 1 to 5 PSI.

We tested the Mini-jector on a '69 Pontiac equipped with a 400 CID V-8, four-speed manual transmission and 3.3 rear axle ratio. Because the pressure regulator is adjustable we made steady-speed test runs and continued to decrease fuel pressure until some improvement was noted in fuel economy. At that point we retained the "maximum economy setting" and ran our full series of steady-speed tests.

Any change from our baseline figures was well within the margin of error of 2%. Actually the deviance was closer to 1% than 2%. At the higher speeds, the difference would simply have to be called "no change."

In making the acceleration runs, the Pontiac ran out of fuel at 65 miles an hour. Fuel starvation was so severe the car would not achieve the goal of 80 MPH although it did so with ease *without* the Mini-jector. In summary, the difference in gas mileage could not be measured since it was well within our 2% margin of error. At the very-low-pressure setting which gave the most change from baseline the car ran out of fuel in the performance test.

Problems caused by excessive pump pressure or carburetors being out of adjustment can be cured by a fuel-pressure regulator in some cases. For instance, off-road race cars often suffer from flooding

In some circumstances, a fuel-pressure regulator is a useful addition—even if it is called a *Mini-jector*.

A relatively simple diaphragm and spring arrangement controls fuel pressure in the *Mini-jector*.

and plug fouling due to excessive fuel in the intake system—caused primarily by the needle and seat and float assembly bouncing around and allowing an excessive amount of fuel in the carb. A fuel-pressure regulator is very beneficial in such cases.

To say a fuel-pressure regulator will increase gas mileage on the highway—much less performance—is to assume the fuel system is not in proper working order to begin with and that the pressure regulator can cure the problems. Sort of like using a ball-peen hammer to kill a fly; the hammer may kill the fly, but it might not be the best way to solve the problem. Additionally it might cause a new problem or two.

SUPER SPARK PLUGS

Often you will see a full-page ad offering you a set of special plugs for your car that are represented to be far superior to any of the standard brands. These are called by a lot of names—such things as "flame injectors," "aircraft-type," "lifetime," and other things.

You have to be guided by common sense. If these plugs really offered large fuel-economy increases, everybody would change to them in about four days. The car makers would choose them as standard equipment and all the big plug makers would start copying the design. Never mind about the patent claims.

In fact, the federal government is now so deeply involved in passing laws about how cars are built and equiped that they would instantly require use of *anything* that improves fuel economy significantly. Nobody has a stronger urge to get on TV and tell you what good things they are doing for you than the politicians and government experts.

The facts are, the world generally ignores the special spark plugs and uses the major brands backed by research and experience. You should too.

THE FRONT END REBUILD GAME

You don't have to be on a trip to be suckered on a front suspension rebuild. The uninformed can be stripped of $50 to $100 to replace all those "worn parts."

Here's how it works. A mechanic or station attendent gets one or both of the front wheels off the ground with a jack. He may be changing a tire for you, rotating tires, or perhaps you came in complaining the car pulled a little to one side. With the front wheels off the ground, the rest is easy. The con artist grabs a tire and shakes it back and forth. It's loose. He might even get a good resounding metallic clunk! That car is unsafe—the whole front end is worn out. A front-end rebuild is a time-consuming and expensive job—but it's got to be done if you want to keep driving the car.

The facts are, anyone could have grabbed the wheel, shaken it and maybe even gotten that ominous sound the day the car rolled off the assembly line. When the car is jacked up, the weight is off the ball joints in the front suspension. They are loose by design when off the ground. When the weight of the car is on the ball joints, the parts fit perfectly. If ball joints are lubed on a regular basis, they'll last anywhere from 60 to 100,000 miles without replacement. Don't panic if you are told you need a front end rebuild—get a second opinion on the matter from a shop with a good reputation for honesty. You could save a lot of money by getting that second opinion.

A better known and better quality brand is the *Filt-O-Reg* unit shown here. It is an adjustable fuel-pressure regulator with the adjustments marked in pounds of pressure.

The *Filt-O-Reg* innards appear to be of better quality.

ADDITIVES

If you have spent any time at all in auto parts stores or the automotive section of a large store, you are more than just a little aware there are some additives for car engines. At a moment's notice you could name two or three; you have been liberally exposed to them by radio, television, publications and attractive packaging and displays.

You may have several questions—should I use the additives in my car? Do they do any good? How often should I use them? Should I use them when I have engine problems, or should I use them on a regular basis? In the long run, will they save money or cost money? All these are reasonable questions demanding answers that make sense—something the advertising doesn't quite do.

Gas Boosters—There are different types of additives. To start with, let's take a closer look at the "gas booster" type of additive. Normally, the contents of the can are poured directly into the gas tank. The claimed results are: Better mileage; stops stalling and rough idling; prevents carb icing and prevents the fuel line from rusting. Some additives of this type also claim to "restore" horsepower, acceleration and performance.

We visited a parts store and located an 8 ounce can of gas-additive for 69 cents. Directions said to use one can for every 20 gallons of gas. These additives consist of various lubricating and solvent chemicals. The result is a combustible mixture designed to dissolve carbon and other residue that has collected in the carburetor and on the valve stems. Because this additive goes wherever the gas/air mixture goes, sooner or later it all ends up in the combusiton chamber to be burned. The manufacturers claim or strongly imply that before it catches on fire and makes a hasty exit past the exhaust valve stem it will also have time to lubricate the uppermost piston ring and keep it free of carbon and varnish.

The claims for this type of additive need careful reading before you lay the money down. Let's take the claim about the prevention of carb icing and the prevention of gas line rust. Just how often are you apt to face those problems in a lifetime of driving? Carb icing can occur but only under high humidity, low temperature situations—approximately 40°F.

Read the label on this gasoline booster product. Assume it does all those things quite well. How many of them do you need? How many are not accomplished by additives already in the gasoline you buy? Do you want to pay money for this?

One of these oil additives boosts the *power* of your oil. The other one doesn't really say what it does for you.

is the problem temperature. Nearly all passenger cars today have provisions for heating the carburetor inlet air and the base of the carburetor which virtually eliminates the occurrence of ice. Besides those built-in heaters, today's cars have such high underhood temperatures the possibility of icing is remote.

Perhaps a short comment on how the carburetor air is heated is in order. One system, referred to as the Controlled Combustion System (CCS) was designed primarily as an emission-reduction device. Warm air heated by the exhaust manifold is fed into air cleaner so carburetor air is warmed and controlled to approximately 100°F. A thermostat control does this automatically.

Now about the fuel-line rusting problem. This is a super-remote problem if you *never* do anything to prevent it. A stored vehicle is subject to such a problem but a little engine oil mixed in the fuel before storage would be an adequate rust preventive. For many years, gas lines have been constructed of plated steel. Obviously the inside of a gas line would have to be subjected to a lot of moisture before rust could form. One of my friends is currently rebuilding a 1940 domestic car. Before the restoration began, the car had spent 19 years in a field in South Dakota. To complicate the problem for the gas line, the gas tank cap was missing. The gas line was rust free. Marketing slogans are intended to get you to buy. Don't go for it if you really don't need its claimed benefits.

"Stops stalling and rough idling" is another frequent claim for gas booster additives. This is predicated on the assumption the stalling or rough idling is caused by some part of the carb being gummed up and momentarily cutting off the fuel supply to the engine or that a valve is stuck due to residue on the stem. A third possibility is a plug is carboned up and either misfiring or not firing at all. All of these maladies can and do afflict cars.

If your car is so afflicted, an additive of this type can be of benefit. However, before trotting off to buy several cans of additive to cure a problem pull one spark plug and examine it for carbon. If the carbon build up is heavy, then you might opt to invest some money in a new set of plugs. Or at least get a can of quality additive that can be poured *directly* down the carb to get at the problem—the plugs—in short order. There are some quality canned products that you can rely on *when they are needed*. Millions of dollars are spent yearly by motorists who do not have a need.

Claims of better mileage and the restoration of power and acceleration are again based on the *assumption* a valve or piston ring is sticking due to carbon and varnish. If the chemicals can dissolve the intruding matter, then power, acceleration and gas mileage can be restored to the level they were before the crud got in the way. Do not read more into the claim of "better mileage, power and acceleration" than is there. If the engine is clean internally then the additive will be of little or no benefit. If the engine is dirty on the inside and the additive does eliminate a problem by removal of residue, performance may be increased. But it will not be increased above the level it was before the residue got in the way. Voodoo and black magic are not sold in a can. Some quality products, when properly applied, do eliminate some problems.

Before buying the "gas booster" type of additive, you should know modern gasoline already contains additives designed to do *everything* the gas booster additive does; thus everytime your gas tank is filled, a new load of cleaning additives is added at no extra charge.

To reiterate, gas-booster additives are often capable of living up to their claims. Some of the claims are rather limited and based on rare occurrences such as carb icing and gas line rusting. The other claims assume the engine is malfunctioning due to carbon buildup or other residue within the engine which can be dissolved by the additive.

Should you use this type additive on a regular basis or just when you have a problem? The argument to use the additives regularly—every tank or every other tankful of gas—is that if the additive will *dissolve* residue, then it will also *prevent* the residue from forming. The argument against this is a look at how much this regular treatment will cost. If you fill up with gas four times a week and use a 70-cent can of additive with each fillup, the additive bill will come to 145 bucks a year! Hardly an economical move. In five years the bill would come to more than $700—enough for a new or rebuilt engine.

Oil-containers like this are disappearing from the scene but you still see them occasionally, lined up and waiting for the passing motorist. I would never buy oil in one of these unless I asked for *and paid for* the cheapest grade of oil available.

Many engines have passed the 100,000 mile mark on pump fuel with no additives.

Oil Boosters—The second type of additive in common usage is the "oil booster—viscosity builder." This is added to the engine oil. The claims are the additive will reduce oil consumption, quiet the engine, restore compression and improve engine efficiency. Whereas the gasoline additive is very thin and slightly oily, the oil additive is often very thick—to the point it is almost impossible to pour on a cold morning.

Oil-additive claims assume the engine they are being used in is practically worn out. If the engine in your car is burning one or two quarts of oil every three or four hundred miles and is exhibiting an ominous sounding knock when going up a steep hill or when pulling away from a stop, then an oil additive may help. This type of additive thickens the oil. If you are currently using 30-weight oil in your clunker, a can of the oil additive can bring the crankcase oil up to the equivalent of 40-weight oil. Ring seal may be improved, in some cases the knock will go away and for awhile the car will seem to run better. If the spark plugs have been fouling with annoying regularity, then this condition may even be improved.

The super-thick oil additives are simply viscosity boosters which make the oil thicker. If you must nurse a sick, worn-out engine along, you are probably doing so because you don't have the money to get it fixed correctly. In this case you can come out dollars ahead by avoiding the rather expensive oil additives and going directly to a thicker oil. The same effect can be achieved at far less cost.

Also remember if you are trying to nurse a tired engine a few more miles; treat it like an old tired horse. Keep the pace down, do not flail it. If all the pieces are intact, a very tired engine can be coaxed for thousands of miles.

ECONOMY CAMS

A camshaft opens and closes the valves to let the air/fuel mixture in and the exhaust gas out. Camshafts in older passenger cars are designed to give reasonable idle, economy and performance. And, in recent years they have also been designed to help control emissions—at a sacrifice of performance and economy.

Those who design and build camshafts

for racing offer many configurations—each for a specific application. So also with the crop of so-called economy cams. Speaking generally about economy cams, the route they take to improving gas mileage is to shorten the duration or the amount of time the valve is open.

If you get an economy camshaft exactly tailored to your engine, your vehicle and driving habits, you could see an improvement in gas mileage of 1 to 4 MPG with no other changes. Before getting into our test results, there are a couple of catches you should be aware of. Notice we said, "camshaft *exactly* tailored to . . . ."—that's one catch. *If* your cam supplier has an economy cam exactly suited for your application, *if* you haven't fudged on the requirements, and *if* you can keep your fat foot out of the throttle when the light turns green, you can get some increase in gas mileage and low speed performance. The second catch is that few people realize the amount of money, time, effort and skill needed to install any camshaft in most engines correctly. On an overhead cam engine like a Datsun the job is fairly simple involving a minimum amount of time and tools but more than casual knowledge of the engine. On a modern V-8-powered full-size car the job can entail a full day's labor for a guy that doesn't waste any time.

When I installed an economy cam on our test truck, the distributor, intake manifold, radiator and grille had to be removed. In addition the Freon had to be discharged from the air-conditioning system because the condenser had to be removed. To do the job correctly you should install a new timing chain and new valve lifters. This plus the gaskets and miscellaneous items needed for the job will run the cost up to around $125—and that's not counting any of your labor. A quick check of several reputable garages indicated they would charge $80 to $100 for installing a cam in most late model V-8-powered cars. At an average price of 60¢ per gallon, $125 will buy you 208 gallons or gasoline, or the total of $200 spent to have someone install the cam for you will buy 333 gallons. Something to consider when you weigh how the job might possibly turn out.

To find out just what an economy camshaft would do, we asked one of the popu-

Swapping your stock camshaft for an economy grind is not a simple proposition. Be prepared to strip it down to the bare essentials.

The lobe on the economy cam at right is shaped differently than the stock cam at left. Our test vehicle did not benefit by the exchange.

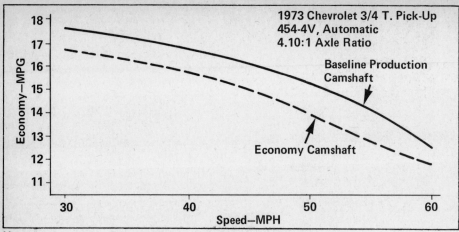

Here are the results. Fewer miles per gallon at all tested road speeds.

lar cam grinders to furnish us with an economy grind for our 1973 Chevrolet 3/4-ton crew-cab pickup. This test truck was equipped with the 454 CID engine with four-barrel carb, automatic transmission and a 4.10:1 axle ratio. We ran a baseline economy and performance test on the truck and then proceeded to install the new cam.

The steady-speed testing of the economy camshaft is documented testimony to the "all is not gold that glitters" thesis. At the lower speeds the economy cam lost nearly 1 MPG—far out of our 2% error range. At a steady 60 miles an hour the cam was losing about 1/2 MPG—all that work and money produced a substantial loss of economy. The camshaft was installed as per the manufacturer's instructions—which simply means the cam was not suitable for the vehicle. With this much loss in gas mileage we could only hope for a considerable gain in performance. Such was not the case.

Performance testing (0 to 30 MPH acceleration) showed the economy cam to be just about 3/10 of a second slower. Gas consumption changed only about 2/10,000 of a gallon average on all the acceleration runs. That's less than one CC—less than 1/2 teaspoon.

We feel the slow-turning engine in the test vehicle coupled with the weight of 6060 lbs. worked against the nearly stock duration and higher than stock lift to kill low end torque and economy. This same economy camshaft would have probably worked quite well in a lighter car with a stick-shift transmission and higher numerical axle. Chances are an aftermarket cam with substantially less duration and lift would have made an improvement in the low-end torque and gas mileage of our test truck. But that is not what the cam grinder selected—and that is not what we tested. We were sold the wrong cam for the application—it could happen to you!

HOW WE RAN THE TESTS

Every gadget has a fan club and people who supply testimonials for the ads. Nobody can know how the manufacturer of an economy-improving gimmick actually feels about his product. However, I think the average citizens who buy them and then testify that they work well are probably convinced.

Unfortunately, a lot of people get themselves convinced through wishful thinking or allow themselves to be unduly influenced by the fact that they paid money for the gadget and therefore it must work. The average person doesn't have the time or know-how to make a careful test and to start out by measuring economy before the miracle gadget is installed.

So there will be some people who argue with my conclusions. All I can say is that I have been a professional automotive engineer and tester for many years as a salaried employee of a major car manufacturer, as a consultant, and as an independent business operator. I used the same methods in testing devices for this book as I use and have always used professionally for the automobile industry. I believe my results are accurate.

There is no magic to good test procedures. We used a $2,500 Autotronics fuel meter to measure gasoline consumption. We even ran the fuel line through a special ice-filled cool can to make absolutely sure there was not a hint of vapor lock which can alter the accuracy of the meter.

At the beginning of each test procedure, we go out to a level measured mile and make baseline runs to establish performance before changing anything about the car. There is a crew of two people. The driver starts and stops the fuel meter, watches the speedometer to keep a steady speed and repeats the runs until the test data repeats consistently during several runs.

The observer measures the time it takes to run the mile, records the fuel flow meter readings and resets the meter for the next run. Most test procedures require runs at 30, 40, 50 and 60 MPH. There were other kinds of tests also, depending on what was being tested and the purpose of the test.

Runs were made in both directions to compensate for slight changes in road elevation or surface winds. When we had consistent test data at one speed, the entire baseline test procedure was repeated at the next speed.

Then a change is made (let's say it was an ignition change) and the test procedure repeated all over again to find the effect of the different ignition. In this manner, tests can be compared with the baseline runs so there is no doubt what affect the change had on the amount of fuel being consumed. This was time-consuming. Testing always is—when you want to be sure of the data.

Acceleration runs were done in much the same manner using a half-mile course. At the end of each run, the amount of fuel used was recorded, the speed at the end of the measured 1/2 mile, and the time it took to get there. One run would be made traveling east, the next one west and so on until the fuel-readout figures would repeat. Our requirement that the fuel readout, time and speed figures repeat is the key to the accuracy of the test data.

6 BUYING A CAR FOR ECONOMY

Each year millions of people face the prospect of buying a new chariot for the family or business. This purchase is one you may have to live with for a number of years, so prudent selection can mean the difference between satisfaction and an empty pocketbook. You should fight the impulse to buy the first car that looks good. If you can order a new car with the exact options needed, the wait of 4 or 6 weeks will be more than compensated for by getting a car tailored for the type and style of driving to be done. Buying a used car provides a little more of a challenge in that the only way to get the options you want is to *SHOP*.

WEIGHT

Many factors affect fuel economy and vehicle weight hurts most. For example, compacts of today often weigh 3000 pounds or more. The hot tip is buy only as much car as you need. If you don't really need a Monte Carlo for driving to work; buy a Vega. If you don't need an LTD Ford buy a Maverick or Pinto. The make is not as important as the size. You will save on the original purchase and save many dollars in gasoline alone over the life of the car. This is ECONOMY in its truest sense.

How does weight affect the inertia and rolling friction of a car? If it takes a certain rate of acceleration to keep up with traffic in a certain situation, the amount of force required is directly proportional to the weight of the car. For instance, if a 3000-pound car takes a certain amount of force to accelerate to 60 MPH from a stop, a 6000-pound car takes twice as much. The heavy car uses more gasoline.

AERODYNAMICS

Automotive aerodynamics is a subject that could not be covered in ten books this size, but there is one simple principle easily applied to this discussion: The force required to overcome the resistance of the air against the front of the car is

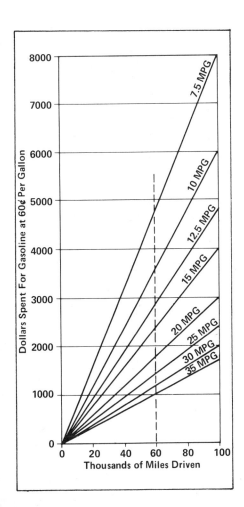

BUYING FOR GAS MILEAGE? HOW MUCH CAN YOU SAVE?

People are shopping for fuel economy these days to reduce the cost of buying gasoline. If you switch to a car that uses 10% less fuel, you should save 10% on fuel costs, which is a worthwhile saving.

Unless you spend more to get the saving than you save. Would you spend $2,000 to save $675? Probably not. Before buying on a miles-per-gallon basis, use this chart to figure your actual saving in dollars. Then figure the impact on your wallet or bank account.

The chart shows thousands of miles along the bottom line. You can use this as thousands of miles per year, or the total mileage you expect to put on the car before trading again.

The vertical line is marked off in total dollars spent for gasoline, assuming a cost of 60¢ per gallon. If you pay more or less, you can adjust the figures accordingly.

The angled lines on the chart are for cars with different fuel consumption rates, shown in miles-per-gallon or MPG. Let's make a quick comparison based on operating a car for 100,000 miles, which is the highest number on the miles-driven scale. The line for 20 MPG shows a fuel cost of $3,000 for that many miles driven. If you switch to a car that gets 30 MPG, the chart shows a total fuel cost of $2,000—a saving of $1,000 over 100,000 miles of driving. Now you can compare the other aspects of the 30 MPG car against the 20 MPG car and decide if the overall deal is worth it to you.

You can use this chart to make comparisons at any miles-driven figure you want, by drawing some lines on the chart. The dotted lines are an example, based on 60,000 miles. A car that gets 12.5 MPG will cost $2,900 in fuel. A car that gets 20 MPG will cost about $1,800 in fuel. The difference is $1,100 and that's another input into your decision.

Don't base any car-buying decisions entirely on fuel economy and don't believe everything you hear about MPG. My neighbor wants to sell me his giant station wagon with power everything. He says it gets about 19 MPG around town and a lot more on the open road. I tell him a car like that is priceless.

93

Big fat car.

Trim little car.

Look how much of the big fat car overhangs the profile of the trim little car. There is a big difference in frontal area—and that costs money in fuel consumption.

directly proportional to the frontal area of the car. As an example of how aerodynamics affect fuel consumption let's consider the 3000-pound and 6000-pound cars again. Assume the two cars are roughly the same shape so we can consider the effects of frontal area without getting into other aerodynamic considerations. The 3000-pound car has a frontal area of 24 square feet. The heavier car has a frontal area of 35 square feet. Because the force required to overcome the resistance of the air is proportional to the frontal areas, the additional force required to keep the larger car at the same speed as the smaller one can be calculated as follows:

$$\frac{(35 \text{ sq. ft.} - 24 \text{ sq. ft.})}{24 \text{ sq. ft.}} \times 100\% = 45.83\%$$

The 6000-pound car would require 45.83% more force and approximately 45.83% more gasoline energy to overcome aerodynamic drag than required for the 3000-pound car at the same speed.

What all these reductions in fuel economy for the big car add up to is just what you observe when driving it. *Smaller cars provide much better fuel economy under all reasonable driving conditions than large cars. The car buyer can expect about the same percentage decrease in his fuel economy as the increase in weight when he decides to buy the larger car.* This means if you go from a 2000-pound economy car getting 26 MPG to a 4000-pound intermediate size car, the fuel economy should decrease to 13 MPG. From an intermediate-size car to a 6000-pounder, the fuel economy will drop from 13 MPG to 8.7 MPG. These calculated economies compare very closely with the actual data published by the Environmental Protection Agency (EPA) and by the reliable testing magazines like *Consumer Reports.* We all are aware some differences in mileage will result from one manufacturer doing his homework a bit better or choosing the compromises that are least detrimental to fuel economy.

ENGINES

When you have decided what size your car should be to meet your needs, the next consideration should be which standard or optional engine to put in the car. If you buy a car with a large engine you may not get good fuel economy. What most people don't know is they stand to lose as much or more by under-powering as by over-powering. Due to emission standards there have been changes in late-model engines such as low compression, inefficient camshafts, poor combustion chambers, and carburetor metering too lean for some modes of driving and too rich for others. This means the buyer cannot always rely on past experience with older-model engines to determine which engine to buy.

Most drivers have ingrained driving habits that are extremely hard to change. This means engine size and power should be selected for the driver and his or her driving conditions. A driver with conservative habits who can limit himself to moderate accelerations and is lucky enough to drive in the type of traffic where he won't cause a hazard driving that way can use a six-cylinder or small V-8 for any car except the larger ones. On the other hand, if the driver is more aggressive or traffic demands harder driving, a more powerful engine runs easier and probably will get better fuel economy. The only way to assure yourself an engine adequate for your needs is to test drive similar cars with the standard and optional engines and decide which package is best. Pick the smallest engine that meets your performance requirements in most situations. If the engine is too small, you'll know it because heavy throttle will be the rule, not the exception, and it will be necessary to downshift the transmission or hold the transmission in the lower gears for longer periods of time just to make the car accelerate the way you want it to.

The old myth about 2-barrel carburetors being more (or less) economical than 4-barrels should also be covered. The four-barrel has four air passages: Small primary air passages give good efficiency at lower engine speeds and the large secondary air passages allow the engine to produce maximum power when it's required. A large engine with a four-barrel carburetor may have more economy potential than with a two-barrel. There is no firm rule or way to tell which—a four barrel or two barrel—will be best for you. The engine, axle ratio, weight of vehicle and other factors all influence economy.

In the 1975 Union 76 Economy Tests, a Cadillac with a 500 cubic-inch engine got better city mileage than a Mercury Comet with a 250 cubic-inch six-cylinder and a single-barrel carburetor.

AXLE RATIOS

Axle ratios are a controversial subject because how the axle ratio affects a particular car's economy depends on how and where the car is driven. Determining which axle ratio is best for your car is done in much the same manner as picking the right engine. The engine should turn as slowly as possible to conserve fuel while cruising without turning so slow that heavy throttle and frequent downshifts are required in normal traffic. If you have to run in the lower gears a lot, the car may even show a decrease in fuel economy if the car is equipped with a low numerical "economy" gear. The same car, on the other hand, would show a definite increase in mileage when driven primarily on open roads and highways.

Federal emission regulations require manufacturers to certify each engine/transmission/axle package they wish to sell. This expensive process includes various tests as well as the 50,000-mile durability test. For this reason, fewer axle-ratio options are available from the factories. Usually the standard axle is the one to buy for economy; it will be a lower numerical ratio to keep the engine running slower. This gear will be a compromise between the best city driving axle ratio and the best highway axle ratio and will probably be the best you can buy for all-around use. An optional axle available in a "trailer-towing package" is intended for cars towing a trailer and may benefit an under-powered car or one driven primarily in heavy traffic.

AUTOMATIC OR STICKSHIFT?

It is rare to find a large car with anything but an automatic transmission and a large percentage of the smaller cars also have them. Two generations of American drivers have grown-up with the automatic and most never developed the skills necessary to handle a manual. But, due to the inherent economy advantages of the manual, it is time to change that.

Because there is no direct mechanical link inside the automatic transmission, there is always some loss due to slippage. A standard clutch has 100% efficiency because it doesn't slip. But, automatic-transmission efficiency varies from 0% when the car is in gear but not moving to a maximum of about 95% at high speeds.

The economy advantages of the manual transmission are due to the fact it is an almost 100% efficient method of transmitting engine power to the rear wheels. The automatic on the other hand loses from 5% to 20% of the available

An automatic transmission normally costs you more at time of purchase and more during the life of the car. However, the penalty is relatively small on big cars compared to small cars. If you're gonna live it up in a highway barge, you may as well have the convenience of automatic.

engine power under various driving conditions. The variation is due to the differences between automatics found in various sizes and types of cars. A smaller car with a small engine will lose about 20% fuel economy compared to an identical manual-transmissioned unit. Larger cars with powerful engines may lose only about 5% fuel economy. Some will be worse.

An automatic transmission usually adds $400 to $600 to the base price of the car. This is a reasonable price considering the complexity of the automatic over the manual, but the cost to you doesn't stop when you lay the money down for the car. If the car is a large one and your economy drops from 10 MPG to 9.5 MPG the approximate additional cost for fuel in 100,000 miles will be about $300 if gas prices don't increase more than they already have. The smaller car in the 20 MPG class which drops to 16 MPG with an automatic will cost about $700 more than the same small car with a manual transmission to run 100,000 miles.

If you have already decided to buy the large car and have resigned yourself to the extreme fuel-economy penalty associated with it, the automatic doesn't hurt that much. Anyway, the large cars need the automatic more than the small cars do.

If you are buying a smaller car because of its lower price and better fuel economy, the automatic *may be* a poor choice, but it's not a clear-cut decision.

Even though an automatic does cost more to buy and more to operate, it is much more convenient to use in heavy traffic with a lot of stop and go. If the members of your family learned to drive cars with automatic transmissions, they may not like to use a manual transmission and may never learn to use it properly.

An automatic transmission is easier on other mechanical parts of the car—drive line and tires for sure—if compared to a manual transmission used roughly and used to make jackrabbit starts.

Automatic transmissions don't give as much engine braking when slowing down so you wear out the brakes faster with an automatic, particularly in traffic and city driving with a lot of stops.

A manual transmission does give better engine braking when you take your foot off the gas, and as you near a full stop you disengage the clutch so the engine doesn't try to keep the car rolling. Therefore

Add-on air conditioners cost less to buy and less to operate because they recycle the air inside the car.

brakes can last longer with a properly-used manual transmission.

If the people who drive your car have the skill necessary to use a manual, or if they are willing to learn, it may be a good buy, provided you drive your car with manual transmission about as smoothly as an automatic transmission does automatically. If you jerk the car, slip the clutch, and hot dog the car around town, a manual transimission may cost about as much as an automatic in the long run.

If you are serious about saving money, there is a savings potential in using a car with manual transmission, but you have to drive conservatively to earn the savings.

AIR CONDITIONING

Air conditioning in automobiles and trucks has gained in popularity over the last ten years. Everyone knows it is an expensive option to buy originally, but very few people realize what it costs in fuel economy and maintenance during the life of the car. It's nice to be comfortable, but the cost of comfort is often sizable.

A car is difficult to cool to begin with. The large amount of glass area required to give the driver adequate visibility also lets in a large amount of solar heat. The car roof and body are not well-insulated. You have probably experienced the situation where a car parked in the sun on a cold day will be nice and warm inside. This may be a good thing in the winter, but the same thing occurs in the summer too. It is typical to find interior temperatures of 130° to 160°F. inside a closed car on a hot summer day. This can even be worse on a dark colored car or one with a dark colored interior.

The temperature inside the car may go down because the windows are opened or the air conditioner (AC) is turned on, but heat is still coming in. Other sources of heat such as engine heat through the firewall and drive-line tunnel, exhaust heat through the floor and heat from the road all contribute to the interior temperature of the car. What does it take to cool an averaged sized car? About as much air-conditioning capacity as a medium sized house!

How much does it cost to run the AC on your car? It depends on a lot of factors. The initial price of AC and maintenance costs are fairly sizable. Dealer list prices vary from about $400 on small and medium sized cars to almost $600 for the fancy "temperature controlled" or

Factory air like this may mix hot air from the heater with cool air from the air conditioner to give the temperature you want. For operating economy, set it for full cold and control car temperature with the blower speed control.

Factory air with a temperature-control dial is normally most expensive to buy and operate.

"climate controlled" systems available on most large cars and some of the smaller ones. Maintenance will probably average about $25 per year for belts, hoses, Freon, compressor clutches and compressor repairs if you keep the car for 5 years. You can pretty well count on a total cost of $525 to $725 for the factory AC system no matter how much or how little you use it. And that does not include the additional cost of gasoline to drive the air conditioner unit.

Fuel economy losses on an air conditioned car depend on driving conditions, outside temperature and humidity and the type of air conditioner you have. There are three different air conditioning systems available which you can buy from the factory or after-market suppliers.

The first and most efficient from a fuel economy standpoint is the recirculating system. This is what most of the after-market suppliers sell and is also the one used by Detroit until the mid-60's. The recirculating system takes the air at floor level, cools it and blows it into the interior. This air circulates through the interior and comes back to the AC unit partially cooled. The air is run through again and the process starts all over. Because the same air is only cooled once and then kept cool, the air conditioning costs decrease dramatically. One other important consideration in favor of after-market full recirculating units is the initial price; usually in the range of $300 to $350 installed. Since this type of unit requires less compressor capacity and also because there is less gadgetry such as vacuum-operated valves in the unit, the selling price to you is usually much less than the factory air system.

The second type of AC system is the one most commonly found in factory installations. This unit is combined with the "flow through ventilation" system and is a non-recirculating system. Outside air is taken in at the base of the windshield, cooled in the AC unit and blown into the interior. The cooled air flows through the car and out the vents at the back.

Because the air must be cooled from outside temperature which can reach over 130° down to 50° or 55°F. at the outlet, the cooling capacity has to be much larger than a recirculating unit and the amount of fuel energy required to

97

cool this air is also much larger.

Factory air systems have a sliding lever control for the temperature of the outlet air in the car. This lever also controls the heater temperature in cold weather. Think about it! What is happening when you adjust this lever for less cooling is the AC unit is mixing hot air from the heater with cold air from the air conditioner to get the temperature you have dialed up. You haven't decreased your AC load at all! If you have a system like this, control the cooling with fan speed. Keep it on full cool and turn the blower speed higher or lower as needed.

Some factory air systems have a feature by which you can select "inside" or "outside" air for air conditioning purposes. This is a good feature because you can select "inside" air for a partially recirculating system and some fuel economy gains. The "outside" air selection can be made if there is a cigar puffer in the car and you want a steady stream of fresh air moving at all times. Most of the manufacturers have gotten away from this set-up in recent years—we assume for a cost savings in manufacture.

The third type of system is the temperature-controlled air conditioner. This unit allows the driver to set a temperature with a dial mounted on the dash. The AC system then either heats or cools the air to maintain that temperature in the interior. This type of unit retains all the poor economy features of the standard system and adds a few of its own. When the automatic feature is engaged, cold air from the AC is mixed with hot air from the heater to regulate temperature. Since this air mixing is done continuously, there are air conditioning loads and fuel economy losses at all times; even in the fall and winter months! Enough said.

IGNITION SYSTEMS

In 1973 Chrysler added a special electronic ignition system to most of the cars in their line. This consisted of a magnetic impulse distributor and a "black box" that eliminates conventional mechanical points in the ignition system. Ford has a very similar system that can be purchased as an option for about $40. GM has offered a special system since 1971 on selected models. This unit is different from Chrysler's and Ford's in that it is a combined magnetic impulse distributor-electronic ignition-high voltage coil; all in a package as small as a normal distributor. It has been available for most V-8's in all of GM's lines and it sells for $65. Do I recommend these systems? Without qualification I do!

Factory-installed electronic ignition is a good buy. Learn to identify it on the brand you are buying and check under the hood, just to be sure.

The super ignition systems available now are a direct offshoot of the work that has been done for racers over the last 20 years. A race car needs two things from its ignition system: A spark of high enough *voltage* and long enough *duration* to fire the cylinders under *all* conditions, and freedom from maintenance during a race. These virtues for race cars are also virtues for street models and here's where $40 to $65 spent on an ignition will mean economy in the long run.

The average car with conventional ignition run for 100,000 miles will require about 8 ignition tune-ups during this period. Two things happen to make tune-ups inevitable; the spark plugs wear out which increase the amount of voltage required to fire them. The points get burned or pitted which decreases the current to the coil and decreases the output voltage to the sparkplugs. How's that for a pair of conflicting conditions? Just as the plugs require more voltage to fire, the ignition voltage is going down because the points are burning.

The bill for a tune-up on a V-8 with conventional ignition should be pretty close to the following:

plugs	$10.00
points and condenser	5.00
labor	20.00
	$35.00

Figuring eight tune-ups in the life of the car, the total cost of ignition maintenance would be close to $280. If you skip some of the tune-ups you'll probably pay for them in reduced gas mileage.

For any of the factory electronic systems you can eliminate the points and condenser from the parts list as well as the labor required to install them and set the timing. Assuming no increase in spark plug life, the ignition maintenance bill goes to:

plugs	$10.00
labor	10.00
per tune-up	$20.00

This is only $160 over the life of the vehicle and a savings of $120 over a standard ignition. Not a bad return on a $65 investment. If you replace the plugs yourself, the saving is even more.

We assumed for the maintenance summary above that spark plug life would not be increased by these ignitions, but it actually is. These systems don't significantly increase the output voltage of the coil, but they *do not* allow the voltage to deteriorate over a period of miles and they maintain full voltage at the plug at all times. This means the plugs can accumulate more combustion deposits before a cylinder will misfire.

There is one drawback to these factory super systems; if they quit, you're walking. Solid-state electronics have come a long way to be durable enough to survive in the underhood environment, but like anything else, they can fail. These ignitions are still pretty rare, so don't expect the corner gas station mechanic to understand it, let alone have parts for it. There shouldn't be much worry, however because if the unit makes it through the waranty period, the chances are very good it will last as long as the car does.

AUTOMATIC SPEED CONTROL

Automatic speed controls have been around for a long time. These devices allow the driver to push a button and the car will automatically maintain speed. Factory price for a speed control varies in the $65 to $85 range and they are available from several after-market suppliers for roughly twice that amount.

Speed controls have some advantages and some disadvantages. The biggest advantage is probably the convenience of being able to push a button and take your foot off the gas on a long trip. This could be important to you in these days of double or triple throttle-return springs on carburetors. Throttle effort has increased substantially since the Federal government dictated the use of multiple springs as a safety measure.

Fuel economy increases are also a plus for the speed control in *some* instances. If the driver is the type that is constantly accelerating and decelerating due to lack of concentration or coordination, the speed control can mean substantial fuel savings. For a good, steady driver, however, there will be little or no gain in fuel economy.

On the minus side are several factors. The average driver will not use the speed control enough to allow it to pay for itself in gasoline savings. In city or busy freeway traffic where the average driver logs most of his miles, using the speed control is inconvenient and can be downright dangerous because the speed control maintains speed until the instant the brake is pushed. It can take you deeper into a panic situation before you can react.

There is one facet of economy driving where the speed control can actually decrease your fuel economy; mountain driving. The speed control is designed to maintain a set speed and if it takes a wide open throttle to make the car go up the hill at that set speed, the control will pull the carburetor wide open.

In Chapter 2 of this book there are techniques which will allow you to consistently match or beat the economy of a speed control. There is no way an automatic controller can get more economy than a knowledgeable driver with a vacuum gauge and a steady foot. It can give that steady foot a rest and this may be worth more than the purchase price to some people. Consider it. Automatic speed controls do what they are designed to do, but they are not economy miracles.

TIRES

There are four factors which determine how long a tire will last on a car: size, construction, (bias, bias-belted, radial, etc.), tread design and durometer rating or hardness of the rubber. Most of these factors are covered in Chapter 4, but I'll recap here to help you decide which tire to order with your new car.

Usually, the standard tire on a car will be of bias or bias-belted construction and sufficiently large to handle the rated load of the vehicle. These tires can be expected to give you the best ride the car is capable of, decent handling and braking performance; and will last about 1 1/2 times the warranty mileage when driven conservatively. Can you select an optional tire that will mean money in your pocket over the life of the tire?

Because you have no control over the tread design and durometer rating of the tires sold to you by the factory, your only options will be size and construction. Tire construction has had a lot of publicity over the last few years because the radial tire manufacturers are actively trying to take a chunk of the overall tire market away from the traditional bias-ply tire. The radial does have several advantages over a conventional bias ply. Acceleration, braking and handling performance are better under most road conditions. Radials run cooler due to reduced sidewall friction and they can ride just a bit better than a similar steel-belted bias-ply tire, but usually not as good as the standard bias-ply. Theoretically, the radial should have superior fuel economy and tire life over conventional tires, but we have found little to substantiate this in our tests or those run by other engineers outside the tire industry. Apparently, since the radial has a performance image to maintain, the tread patterns and softer rubber compounds required to make the tire "perform" have negated part or all of the theoretical economy advantages.

Tire size is another option when buying a new car. A larger tire has a wider tread and a larger "footprint" on the road surface. This gives better handling and braking performance under dry conditions and longer tire wear because the load is distributed over more tread area. The one adverse effect of larger tires is that the tread will lift off the road surface and aquaplane sooner when run on wet roads or slippery surfaces. The larger tread just doesn't push through the water to contact the road surface. Larger tires are usually a very low-cost option when buying a new car and from the economy point of view they are a good buy.

This tire tread pattern has wide grooves to allow water to escape and reduce the likelihood of aquaplaning on the highway. If it rains a lot where you live and you like to drive fast, this can be important.

When talking about wider tires for your car, we are not including the 50, 60, and 70 series super-wide performance tires. Here again the tire companies have "performance image" to maintain. This means bold tread patterns and softer rubber compounds which more than wipe out the advantages of lower tread loads. If you want to compare tires, bigger is better as long as the basic tread design and rubber compound remain the same.

The tire you buy should also depend on the punishment the tire must put up with. The larger bias-ply tire is probably a good buy over the standard tire in all cases. If wet weather driving is common in your area, the more aggressive tread pattern of the radial will cut through the water and give better control. If punctures are a problem, the steel-belted radial or steel-belted bias-ply tire is a good buy. The key to economy tire buying is to buy only as much tire as you need; no more and no less.

WARRANTIES

When you buy a new vehicle, you deserve a car that is mechanically correct. The only economical reason for buying a new car is the peace of mind you get from knowing all components are new and in top condition. But, often there are defects in workmanship and materials which cause problems or inconvenience. This is where the manufacturer's warranty and your dealer comes in. If they choose to, the car can be repaired quickly and completely. But frequently you will get the runaround to some degree. What can you do to eliminate dealer and warranty problems before they happen?

First, shop for a dealer as well as a price. By asking around, you will find dealers with an excellent reputation for warranty and service work. Other dealers will be worse in this respect. Concentrate on the good dealers when shopping price on your new car; but don't ignore the ones with poor servicing reputations. Often if you get a low price quoted by one dealer, you can take it to the one you want to do business with and he will match the lower price.

There is nothing illegal or immoral about playing one dealer against another when shopping for the best price on a car. Keep in mind though, that when you make the final decision you are buying the performance of the dealer as well as the car. If the dealer goes to bat for you on a warranty problem, your chances of getting satisfactory work done are 1000% better than if he directs the service manager to "give 'em the run-around on warranty work."

Second, know the warranty before you buy the car. Most American manufacturers now have basically the same warranty, but there may be differences in some of the details. For instance, American Motors started a plan in 1973 that provided loan cars to people whose new car was being repaired under warranty. If you happen to have no other transportation, this feature should be considered when purchasing a new car.

Some foreign manufacturers have warranties requiring "factory trained" mechanics to perform certain scheduled checks and maintenance. Naturally, the only place to find factory-trained mechanics or authorized service is at the dealership—this gets you back to a dealership whether you want to go or not. In some cases this can cost you $600 to $1000 over a 12,000 mile warranty period for "lubrication and adjustments." Don't think so? Check with someone owning an expensive foreign car. Read the warranty before buying the car!

What recourse do you have if your new car has a problem and the dealer refuses to acknowledge it? Another dealer for the same make car should be contacted first. If this does not bring satisfaction in solving the problem contact the Zone Manager for the brand. The Zone Manager or his representative will contact you—but you had better be prepared to live with the problem awhile since a thirty to forty-five day wait is not uncommon at this level.

DON'T FALL IN LOVE SO EASY

There are some basics to buying a car—most of us know them—all the "do's and don'ts." Unfortunately we get carried away, panic, or fall in love in the dealer's office and we wind up with a car we really don't want, don't need, and at a price we didn't intend to pay.

First, know what you want. Read magazines with road tests, go look at cars, get the sales literature, go for a ride. Just make sure you play it plenty cool around a salesman. NEVER, EVER lose sight of the fact that it is his job to sell cars. Nail down in your mind exactly what you are looking for. Then you should decide exactly how you are going to pay for it. As we point out in the section on financing, if you let a car salesman decide this for you, it will cost you more money. Notice I haven't said anything about price yet? At this point that is of relative unimportance!

First, know what you want. Second, know how you are going to pay for it. Then, armed with your little list of exactly what you want—I said EXACTLY—approach a salesman, lay the facts on him and ask how much that EXACT car will cost.

Now the fun starts. He may ask if you are trading in your old car or buying outright. He may study the list, offer you coffee and doughnuts and go through the motions of looking through a card file or big list to find you that car. Be ready for anything—and be unimpressed. I once went through this little act with a salesman and he pulled a card from his stack in astonishment. His banter was to the effect that he had the car—it had just come in and was on the back lot. He hadn't even seen it. The car was exactly like the one I wanted *except* with a different engine, transmission, seats, console and vinyl roof. Other than that it was just what I wanted!

Take your time, don't drink too much coffee and don't sign ANYTHING! I think it is safe to say you should set aside a full month of spare time—weekends and after work—to buy a car. Why all that time? For most people a car is the second most expensive thing they ever buy—with a house getting the nod for the number one spot on the list of expensive purchases.

Time is on your side if you have allotted some for all this wheeling and dealing. Being decisive is also on your side. When the salesman asks if you are trading or buying outright you pick up a point in the game if you say, "Outright," or "That depends on the price of the new car, and how much you give me for the old one. If we can't get together, I'll sell the old one myself. It's in good shape, and I won't have any problem selling it."

Allowing yourself time for the purchase gives you the chance to go to another dealer to do some shopping. When you get to this stage of the game don't lose sight of your objective. What you want to do is to compare apples to apples, not apples to oranges. A good salesman will ask more questions than you. Chances are pretty good he'll find out you've been to another dealership and you have a price. That's fine—if he doesn't ask, you might even get around to telling him. Either way just keep right on comparing apples to apples—cause for sure he'll want to show you an orange. You want a big engine, he'll show you a small one with a different axle ratio for $500 less.

Be patient, nonchalant, unimpressed—no panic, no romance. Go after the job of buying your new ego wagon with all the passion and emotion you would use to buy $7,000 worth of chicken feathers.

When you find the car you want, don't let on. Just look at it like you've looked at all the rest of them. It's just a car. Are you sure it's the car you want? Right transmission, axle ratio, body style, trim package—no extras you didn't want? Now, if you already have it settled in your mind how you are going to finance the car, whether you are trading or buying outright ask how much it is *out the door.* You want to know how much it is going to cost you to drive that thing out the door. You want to know every little item—taxes, license, dealer preparation and any other charges. You are not interested in "base price," "your price," "22nd aniversary price," "sticker price" or any variations or rewording of the above. You want to know it all in one crushing blow!

Make him an offer. What? Yeah, the salesman quotes you a price. Write it down. Subtract something from it and offer him X amount of dollars for the car. He'll refuse. Don't panic. Disregard the upcoming cup of coffee; be unimpressed with tears or any figures he may show you.

"I'm not supposed to show these inventory cost figures to any customer, but the boss is not here tonight and I know how much you want that car . . . ." Never mind. Make him the offer; give him your phone number; leave.

Don't sign anything. Bear in mind when you walk out, you may have lost the whole deal. That's a possibility. Before you get home some dude may walk in that showroom with a roll of hundreds and pay the full pop for the car you had your heart set on. If not, the salesman will call you. He might not accept your offer; but he'll call. He might meet you half way, offer to throw in 100 gallons of gas free if you buy at his price, tell you he's stalling another buyer who has the hots for the car. I can't tell you what his story will be, but I can tell you he will call.

A lot of cars are sold by emotion. Any salesman knows that for a period of about 24 hours the normal individual suffers brain fade when standing next to the latest model. A good salesman knows if you withstand 48 hours of your own personal torment then you stand a far better chance of being rational. Therefore he wants to close the deal in a hurry.

At some point in all of this back and forth bantering, the salesman may attempt to get you to sign something. He'll have a name for it, and explain it by saying something along the lines of, "all this does is hold the car until you get back with the money, or until your wife comes down to look at the car. I just can't hold the car without this." Don't sign. When you get ready to buy the car, buy the car. If you lose the car, you lose the car. Start over. Millions of cars are sold in this country every year. Know what you want and what you will settle for. Never, ever lose sight of the fact that it is the salesman's job to sell cars—any car he can.

GAS MILEAGE FIGURES—LEARN TO READ BETWEEN THE LINES

Publishing gas mileage figures has now become *the* way to sell new cars and to sell all sorts of mileage improving devices. Normally, car manufacturers and large dealerships advertise a specific mileage figure—while those selling add-on devices rely on a percentage figure—the old *up-to* game. Although actual mileage figures and percentage increases may be dead accurate, you should know that both forms of "proof" are probably examples of *selective* data reports.

"Actual driving tests of a 1975 Belchfire produced 25.6 MPG over a one-hundred mile course. The test car was in showroom stock condition. Mileage

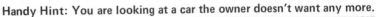
Handy Hint: You are looking at a car the owner doesn't want any more.

figures vary among individual drivers and types of driving."

Does that sound like an ad or two you've seen recently? Of course. Sounds pretty good doesn't it? That's good mileage for a big car like the Belchfire—and over a 100 mile course—boy, that's terrific! You or most any other driver could buy that exact car and never get more than 10 or 11 miles per gallon out of it.

Maybe you should know the Belchfire used for the test had a six-cylinder engine, three speed transmission, 2.73 axle and no air conditioning or power brakes. Maybe you should know the car had 8,000 miles on it when the test started and that it had been tuned and retuned by a really sharp mechanic until he was satisfied the car couldn't deliver better mileage and still be stock. When you think of a 100 mile course, you might think about firing up the car in front of your house, driving 100 miles and ending up back at your house. Don't for one minute think the advertised test was conducted in that manner. Unless the ad flatly states exactly how and where the test was run you can assume someone went to a lot of time and trouble to find a long, predominately downhill one-way course from point A to point B. The test might have been conducted after dark when there is no wind and the air is cool and dense.

Would it surprise you to learn the car was thoroughly warmed up and possibly run up to 80 miles an hour before the driver backed off the throttle—THEN THE TEST BEGAN and never again during the 100 miles did the driver get the Belchfire over 36 miles an hour? You would probably get fantastic mileage figures from your old clunker if you followed the same techniques of preparation and driving.

Sooner or later the Federal Trade Commission will probably lay down some rules concerning the advertising claims of mileage, but until they do the prospective buyer is left to his own imagination as to exactly how and where the test was conducted.

Motor Trend, Car and Driver and *Road and Track* magazines do a good job of testing cars and reporting mileage and performance figures. They are probably as accurate and unbiased as modern journalism permits. In most of the road test stories there is some indication of how the car was driven and some specifics on how the car was equipped. With this knowledge you shouldn't be surprised to see a Belchfire get 10 MPG in a magazine road test article.

The tester most likely drove the car for a thousand miles or so, took it home at night, let a neighbor or friend drive it, let his wife drive it, drove it at a local drag strip to get an acceleration figure or two, and used the air conditioning.

The magazine test car will rarely have more than 5000 miles on it—thus the engine is still fairly tight. If the car is obviously running poorly, the tester may check to see if all the plugs are firing and that the throttle is opening all the way. He may or may not go to the trouble to check timing or tire pressure. On the whole, the average automotive writer will treat the car about like the average owner—but with a more skeptical eye than the guy who carries the payment book.

Ads for new cars and also articles in the press will show the Environmental Protection Agency (EPA) gas consumption figures. This is not simply miles traveled divided by gallons consumed. The EPA is primarily concerned with emissions and they have a complex emission driving schedule that is run on rolls while pollutants are measured. They also have an urban, suburban and highway schedule which is also driven on rolls in a laboratory for the purpose of determining miles per gallon. This is how the ridiculously high mileage figures are arrived at which manufacturers use on the billboard ads and in radio and television commercials. In most cases they are completely irrelevant as to what the car will actually get in miles per gallon. On the rolls there are no head winds, the rolling resistance of tires is different, there is no cold start and warm-up, no cornering forces and thus when you read the published EPA numbers and rely on them to any degree you have been taken in my friend. The car you buy which looks like the one on the billboard will not get that mileage.

Read between the lines—up to 20% of what, on what car, with who driving it, under what conditions. The purpose of advertising copy is to sell—never lose sight of that cold fact.

Actual test data

Even though you cannot expect to duplicate any published gas mileage data unless you run the same car over the same course in the same way, nevertheless you can benefit from tests that are honestly run and impartial. They do tell you which cars do better on that particular test and that can give you a general idea of the performance of various cars in respect to each other.

In the following discussion, cars are grouped together according to size. For each group, mechanical specifications are given along with two sets of gas mileage figures. One is the EPA published data and the other is taken from results of the 1975 Union 76 Economy Tests. In these comparisons EPA data was accrued by driving the vehicles on rolls in a laboratory. Union test vehicles were driven over a road course.

The discussion of the test results may help you decide what type of car you will buy and help you figure what to expect from it.

The specifications listed in Table A represent a number of 1975 vehicles. All are equipped with a manual transmission. You will note the weight is 2400 lbs. or less and the engine size ranges from 85 to 140 cubic inch displacement. Although horsepower is given, it is more because of custom than the actual value it offers in selecting a vehicle for your needs.

The vehicles above should be considered for a second car that one or more members of the family uses to drive back and forth to school, work or short shopping trips. They work out well for single people who simply need wheels rather than a bus or a truck. There are plenty of people who will argue how great they are on trips, but it's my opinion they are primarily to be used on trips of 50 miles or less. There are exceptions to any statement. The Vega has been rated by a number of magazines as being one of the best small cars for highway running. I will have to agree it's not too bad. The 2.92 axle allows it to cruise easily. You should also keep in mind these cars are not meant to be laden with accessories. That's like buying a $50 burro and decking him out with a $2000 western show saddle.

In summary I think the Honda, Datsun,

These and similar cars are the lightweight economy champs.

Recommended as a "personal car" mainly for urban use.

TABLE A
Small Economy Cars with Manual Transmission

	Honda Civic	Datsun B-210	Toyota Corolla	Chevrolet Vega
Rated HP	52	80	88	87
Engine Displacement (cubic inches)	91	85	97	140
Axle Ratio	4.73	3.89	4.10	2.92
Weight	1700	2000	2150	2400

EPA Data—Miles Per Gallon

	Honda Civic	Datsun B-210	Toyota Corolla	Chevrolet Vega
City	28	27	21	22
Highway	38	39	33	29

Union 76 Tests—Miles Per Gallon

	Honda Civic	Datsun B-210	Toyota Corolla	Chevrolet Vega
City	29.6	26.5	20.2	19.6
Suburban	33.0	32.7	26.6	25.8
Highway	31.1	31.3	27.5	26.5

TABLE B
Small Cars with Automatic Transmissions

	Ford Mustang	Ford Pinto	Pontiac Astre	AMC Gremlin	Chevrolet Vega
Rated HP	87	83	87	100	78
Engine Displacement CID	140	140	140	232	140
Axle Ratio	3.55	3.40	2.93	2.73	2.92
Weight	2800	2500	2500	2700	2600

EPA Data—Miles Per Gallon

	Ford Mustang	Ford Pinto	Pontiac Astre	AMC Gremlin	Chevrolet Vega
City	18.0	18.0	19.0	19.0	19.0
Highway	26.0	26.0	28.0	24.0	28.0

Union 76 Tests—Miles Per Gallon

	Ford Mustang	Ford Pinto	Pontiac Astre	AMC Gremlin	Chevrolet Vega
City	14.6	15.6	16.7	15.6	17.4
Suburban	21.0	22.6	24.2	19.8	25.3
Highway	21.8	22.0	24.3	19.4	25.3

Toyota and Vega manual transmission cars referenced in Table A would serve some people very well. Do you make a lot of short trips and does buying gasoline really eat into your budget? Look at the little ones. They are practical and fun to drive for some.

I would suggest you never equip small engine vehicles with an automatic transmission. Table B illustrates some vehicles that are probably at their best when equipped with a manual transmission but because so many people demand an automatic they are shown as such.

These vehicles can serve you well as fuel savers—but drive them carefully. With an automatic you are making a pony pull a plow! A small-engined vehicle can be very disappointing in terms of fuel economy if it is driven aggressively most of the time. Don't buy one if you have a heavy throttle foot.

I don't think these cars should be loaded down with air conditioning and other major accessories. Should you decide this size car is O.K. for trips you take out of town, I suggest choosing one with the lowest numerical axle ratio. Notice that three of the cars are equipped with axle ratios below 3.00. They should be great for engine quietness on the open road at highway speeds.

But don't expect best economy in city driving with a low axle ratio unless you drive very conservatively as is done in the EPA and Union 76 tests. To get brisk performance with a low axle ratio, you have to use heavy throttle a lot of the time. The power system in the carburetor will deliver extra fuel and there goes your economy—out the tail pipe.

The AMC Gremlin listed in Table B is an exception because the engine is nearly 100 cubic inches larger than the others in the group. It can pull the lower axle quite well in city and suburban driving as you can see by the test data.

If you live above 4,000 feet altitude, be sure to test drive a car with small engine and automatic transmission before buying one. The performance and economy is likely to be disappointing.

The next group of vehicles in Table C includes slightly larger cars, with larger engines and more total weight. These cars weigh approximately 12 pounds per cubic inch of displacement—compared to those in the previous group with a weight of about 15 pounds per cubic inch displacement. Emission-controlled engines of today are quite close to each other in the amount of torque and horsepower they obtain from each cubic inch of displacement. When you drop 3 pounds per cubic inch off the vehicle it will produce a bit more performance provided the axle ratio does not change appreciably.

In this group, the average axle ratio is 2.73 and in the previous group it is 3.14. This indicates that city fuel economy is not likely to be especially good. Accelerating performance will be improved because of the reduced vehicle weight per cubic inch of engine displacement.

Because these cars have lower axle ratios, highway economy should be relatively good even though the vehicle size and weight is larger. These cars are large enough to offer reasonable comfort on long highway trips but you pay for it in miles-per-gallon.

Table D introduces still another increase in engine size and although the weight is greater, the pounds per cubic inch of displacement factor is 12 pounds per cubic inch.

You have a bonus at your disposal when you consider the larger displacement engines. They will net you added low end performance over that of small engines. This contributes to the pleasure of driving at lower speeds. The extra displacement will also allow the use of air conditioning and other luxury items that would overtax the capability of vehicles in the previous groups. Although this group of vehicles is not exactly a large family choice for longer trips, it does have the capability of reasonable cruising comfort. Four door models in this group will do a respectable job on occasional trips and serve the average driver well in general use. In my opinion these vehicles are a practical size for many users.

Table E covers even larger vehicles. The weight of these units averages 4200 pounds. Keep in mind you are now having to buy fuel to move two tons everytime you accelerate. Moving weight in city driving is quite costly in terms of fuel being consumed. On the open road you do not need to be quite so concerned with weight. The

Typical compact cars with still more room and some fuel economy penalty.

TABLE C	Ford Maverick	Plymouth Valiant	Buick Skyhawk	AMC Hornet	Chevrolet Monza 2 + 2
Rated HP	72	95	110	100	110
Engine Displacement	250	225	231	232	262
Axle Ratio	3.00:1	2.76:1	2.56:1	2.73:1	2.56:1
Weight	3000	3150	3000	3200	3050

EPA Data—Miles Per Gallon

	Ford Maverick	Plymouth Valiant	Buick Skyhawk	AMC Hornet	Chevrolet Monza 2 + 2
City	16.0	18.0	19.0	18.0	15.0
Highway	21.0	23.0	25.0	24.0	23.0

Union 76 Tests—Miles Per Gallon

City	12.7	13.2	15.2	14.8	14.6
Suburban	16.0	20.3	21.9	19.4	18.7
Highway	16.5	21.3	23.0	19.0	19.8

TABLE D

	Buick Apollo	Chevrolet Camaro	Dodge Dart	Mercury Monarch
Rated HP	110	145	145	129
Engine Displacement	260	350	318	302
Axle Ratio	2.56:1	2.73:1	2.45:1	3.07:1
Weight	3800	3800	3600	3600
EPA Data—Miles Per Gallon				
	Buick Apollo	Chevrolet Camaro	Dodge Dart	Mercury Monarch
City	15.0	13.0	13.0	12.0
Highway	19.0	20.0	20.0	16.0
Union 76 Tests—Miles Per Gallon				
City	13.3	12.2	10.9	12.1
Suburban	17.6	16.0	17.9	15.5
Highway	17.9	17.0	19.8	16.1

Larger engines of this group allow power-consuming accessories such as automatic transmission and air conditioning without big percentage drop in fuel economy.

These 2-ton automobiles are at their best on highway trips. In today's conditions, they may be too much car for daily use in city traffic.

TABLE E

	AMC Matador	Buick Century	Chevrolet Malibu	Chrysler Cordoba	Dodge Coronet	Ford Gran Torino
Rated HP	150	145	145	180	150	154
Engine Displacement	304	350	350	360	318	351
Axle Ratio	3.15:1	2.56:1	2.73:1	2.45:1	2.71:1	3.00:1
Weight	4000	4200	4100	4300	4200	4400
EPA Data—Miles Per Gallon						
	AMC Matador	Buick Century	Chevrolet Malibu	Chrysler Cordoba	Dodge Coronet	Ford Gran Torino
City	13.0	13.0	13.0	11.0	11.0	11.0
Highway	17.0	20.0	18.0	18.0	16.0	16.0
Union 76 Tests—Miles Per Gallon						
City	11.1	12.7	10.6	9.9	9.4	10.7
Suburban	14.4	17.1	15.5	16.3	14.9	14.2
Highway	15.7	18.6	16.9	18.3	15.8	15.3

speed changes are infrequent if you are a reasonably steady driver. It does not take much extra fuel to move 4000 pounds at a steady cruising speed over that of a 3000-pound vehicle if all else is equal. However, the heavier cars generally have a larger frontal area and that causes costly fuel-consuming drag. Most people want the larger vehicles for trips and in most instances I would agree with the merits of the decision. I think the vehicles represented in Table E are the most practical for trips. They are large enough to fill most comfort and space needs and yet are small enough to be fairly economical while cruising.

For short drives with one or two occupants, it's my opinion that this group is bordering on too much weight. Notice the ones getting 9 or 10 MPG in city tests. If you are a charger, that will be even less.

Notice also that three of this group can get pretty respectable fuel economy at an average speed of 55 MPG, as was the case during the Union 76 Economy Tests reported here.

Tables F & G show the heaviest cars available. The average weight in F is 4700 pounds and the average weight of G is 5183. Power to weight ratio is better in these two groups than in any of the previous groups. This means these cars will deliver better performance than the others—but not without costing fuel. To get all of that weight in motion takes energy and that comes in the form of gasoline. These cars are ideally suited for long trips on the open road—but the worst of all groups to drive in town. Cars in either F or G groups can be loaded with any and all accessories without great detriment to their performance. They can also carry heavier loads without greatly upsetting their power to weight ratio.

The message is clear: Use large displacement engines to move large cars and heavy weights long distances. These cars have performance, can handle luxury in the form of more accessories and provide comfort—you pay for all of this in increased vehicle costs and gas consumption. Medium size vehicles with moderate size engines give adequate performance and fuel economy in most all driving modes. They are excellent vehicles to bridge the gap between the very large cars and very small ones in terms of cost, gas consumption, comfort and the ability to

TABLE F

	Oldsmobile Delta 88	Buick Riviera	Chevrolet Impala	Chrysler Newport	Ford Elite	Oldsmobile Toronado
Rated HP	148	205	145	175	148	215
Engine Displacement	350	455	350	400	351	455
Axle Ratio	2.75.1	2.93:1	3.08:1	2.71:1	2.75:1	2.73:1
Weight	4550	4900	4500	4700	4600	4900

EPA Data—Miles Per Gallon

	Oldsmobile Delta 88	Buick Riviera	Chevrolet Impala	Chrysler Newport	Ford Elite	Oldsmobile Toronado
City	14.0	12.0	12.0	11.0	11.0	11.0
Highway	18.0	15.0	18.0	15.0	16.0	16.0

Union 76 Tests—Miles Per Gallon

	Oldsmobile Delta 88	Buick Riviera	Chevrolet Impala	Chrysler Newport	Ford Elite	Oldsmobile Toronado
City	11.8	9.3	10.5	8.6	9.2	10.6
Suburban	15.2	13.3	14.9	13.5	13.9	14.2
Highway	16.1	14.3	16.0	14.6	16.0	15.2

These are big cars, offering luxury and performance but not good economy.

TABLE G

	Buick Electra 225	Cadillac DeVille	Chrysler New Yorker	Lincoln Continental	Oldsmobile 98
Rated HP	205	190	215	206	190
Engine Displacement	455	500	440	460	455
Axle Ratio	2.73:1	2.73:1	2.71:1	2.75:1	2.56:1
Weight	5100	5300	5000	5300	5000

EPA Data—Miles Per Gallon

	Buick Electra 225	Cadillac DeVille	Chrysler New Yorker	Lincoln Continental	Oldsmobile 98
City	11.0	11.0	10.0	10.0	12.0
Highway	15.0	16.0	16.0	15.0	16.0

Union 76 Tests—Miles Per Gallon

	Buick Electra 225	Cadillac DeVille	Chrysler New Yorker	Lincoln Continental	Oldsmobile 98
City	9.2	10.2	7.8	7.9	10.1
Suburban	13.3	13.2	12.3	12.2	14.1
Highway	14.4	14.0	13.0	14.0	15.2

The average weight of this group is nearly 5200 pounds of prestige and luxury. If you remember 'way back in Table A, those little cars are getting twice the gas mileage of these big dudes.

The Union Oil Tests get close to reality because the cars were driven on a road course. Aerodynamics, friction and all the things which affect you are present in these tests. Except one! The cars are fully warmed up for each test so no choke fuel enters in.

pull some of the more popular accessories such as air conditioning. Small displacement, light weight vehicles should be limited to urban driving. They deliver good performance, great fuel mileage and low vehicle cost. They deliver poor performance with heavy loads and they were never intended to be loaded down with accessories.

You may ask why they are sold with air conditioning and other accessories if they can't handle the load well. Manufacturers typically make and sell what people want to buy. Nobody says a small car loaded with accessories will do as well in economy or performance as the same car without the extra burden.

There isn't any single car or type of car that will suit all people for all purposes. What you want from a car and what you are willing to pay are the governing factors that lead to your individual choice of transportation.

The main factor is size—small car, medium size or large—and that's why the preceding tables have grouped the cars in categories. The biggest mistake you can make is to buy in the wrong category for what you want. Do some serious thinking about that before you start shopping—then select a car in the right category.

Here's a more complete listing of cars and the test mileages with no attempt to group them in categories:

	SAE City	SAE Highway	EPA City	EPA Highway
AMC Matador	11.1	15.7	13	15
Buick Century	12.7	18.6	16	24
Buick Century Custom	11.4	17.7	12	19
Buick Electra	9.2	14.4	11	15
Buick LeSabre Custom	10.5	15.3	12	15
Buick Riviera	9.3	14.3	12	15
Cadillac Sedan deVille	10.2	14.0	11	16
Chevrolet Impala	10.5	16.0	12	15
Chevrolet Impala Wagon	7.8	13.0	11	15
Chevrolet Malibu	10.6	16.9	13	18
Chevrolet Monte Carlo	11.2	17.0	13	18
Chrysler Cordoba	9.9	18.3	13	22
Chrysler Newport Custom	8.6	14.6	11	18
Chrysler New Yorker Bro.	7.8	13.0	10	16
Dodge Charger SE	9.7	18.5	13	22
Dodge Coronet Custom	9.4	15.8	13	17
Dodge Monaco	9.7	17.5	12	17
Ford Elite	9.2	16.0	10	16
Ford Gran Torino	10.7	15.3	10	16
Ford LTD Brougham	8.9	15.7	10	16
Ford LTD Wagon	8.7	13.5	10	15
Ford Thunderbird	8.0	14.8	10	15
Lincoln Continental	7.9	14.0	10	15
Mercury Cougar XR-7	9.6	15.4	10	14
Mercury Marquis	9.3	14.2	10	15
Mercury Montego MX	9.3	15.6	10	14
Oldsmobile Cutlass Supreme	12.2	17.2	15	20
Oldsmobile Cutlass S	13.1	17.7	16	21
Oldsmobile Delta 88 Royale	11.8	16.1	13	18
Oldsmobile 98	10.1	15.2	11	15
Oldsmobile Toronado	10.6	15.2	11	16
Plymouth Fury Salon	10.2	16.8	16	21
Plymouth Gran Fury	9.0	16.3	12	17
Pontiac Catalina	10.8	17.4	12	15
Pontiac Grand LeMans	10.1	16.6	13	18
Pontiac Grand Prix	10.0	15.7	12	17
Toyota Corolla	20.2	27.5	21	33
AMC Gremlin 8	15.6	19.4	19	24
AMC Gremlin X 3-speed	16.7	22.1	19	24
AMC Gremlin X Overdrive	15.6	21.8	NA	NA
Buick Apollo V-8	13.3	17.9	15	19
Buick Skylark V-8	14.5	20.9	16	24
Chevrolet Camaro LT V-8	12.2	17	13	20
Chevrolet Nova 6	13.5	18	16	21
Chevrolet Vega 4	17.4	25.6	22	29
Chevrolet Vega Wagon 4	15.9	24.6	21	29
Dodge Dart SE V-8	10.9	19.8	13	20
Ford Granada V-8	11.3	16	12	16
Ford Mustang II 4	14.6	21.8	18	26
Ford Pinto 4	19.6	24.6	18	26
Ford Pinto Runabout 4	15.6	22	18	26
Ford Pinto Wagon 4	14.5	20.6	18	26
Mercury Monarch V-8	12.1	16.1	12	16
Oldsmobile Omega V-8	13.6	18.7	15	19
Plymouth Valiant Brougham V-8	11.1	19.1	13	20
Pontiac Astre SJ4	16.7	24.3	19	28
Pontiac Firebird V-8	11.4	18	13	18
Pontiac Ventura V-8	14.2	18.5	15	19

Used or Abused?

There are many reasons to buy a used car instead of a new one—some very real advantages, some disadvantages. Bargains and bummers—they are all out there. A good buy in the used car market can get you a lot of transportation for a little cash outlay. It can also buy you a lot more performance and overall economy than you can find in a new car of the same size. Depending on make, model, year, and body style it is possible to buy more interior room and more quality of construction in a used car than a new one. Considering the number of used cars for sale in this country on any given day a buyer has a choice of several million used cars.

Before rushing out the door with checkbook in hand you should be aware your checkbook is the only limit to how much trouble you can buy. Knowledge and time are even more in your favor in shopping for a used car than a new one. Why? Simply because it takes more knowledge to appraise something that's been used—or abused—than it does to look at something new. And when you really get down to the nitty gritty of it, it takes one heck of a lot longer.

You can either be very specific about the car you want to buy used, or the amount of money you want to spend. Establish your goals. You want a dependable used car to get you back and forth to work for one year. You have $1000 to spend and you don't care if it's a sports car or a station wagon. That's a different situation from the guy that has $1000 to spend, needs a large sedan or station wagon to haul all of his kids and must make it last just as long as possible with a minimum of money flowing out of the billfold. Make no mistake about it—there are two different cases here. Where do you stand? Create your own case. Establish your goals—your limits. Sometimes it helps to write it down—so you don't forget, so you won't be misled.

There is a lot of difference between looking at anything and everything for $1000 or less and being very selective in

A used car is a used car, whether offered by the car dealer who just got it from a little old lady—or offered by the little old gal herself.

shopping for only large sedans and station wagons that cost $1000 or less.

Every case must be considered separately, but I would put some effort into finding the right car owned by an individual before turning to a used car dealer. You don't want the sales pitch or the pressure. You want time with the car and generally you can get more of that when you buy from an individual.

Always inspect a car in broad daylight or a well lighted garage. Stand first to the front and then to the rear—a little bit to the side so you can view along the side of the body. Do the panels line up? They should come pretty close. Is a panel wavy? If you are not sure, look at the other side and see if they look alike.

What you are looking for is any evidence the car was wrecked. An inner fender panel can show signs of welding or hammering but the outer fender can be in pretty fair shape. Has the car been repainted? Look closely at emblems and other pieces of trim that show signs of paint overspray. If you do find evidence the car has been hit hard, work on it like a dog on a big bone.

A wrecked car can be fixed. A torn fender, dented door or crushed bumper is no big deal. Parts can be fixed or replaced, but when you get a vehicle that has really been hit hard and then repaired, all sorts of problems can crop up. Frames can be twisted, suspension members damaged, engine mounts torn loose, driveshafts shoved up into the transmission tailshaft and the list goes on.

A car wrecked hard can be dangerous to drive and it can cause endless trouble. It doesn't have to be that way—but it can be. Realize that and be on guard for a car that has been hit hard. Tricks at hiding evidence of a bad wreck are normally more refined on a used car lot—another reason to start your shopping for a used car with an individual owner.

After walking around a car a couple of times to generally size it up you can go to work trying to determine its condition.

Are body panels rusted? Do ALL of the doors open easily? Do all of the windows go down easily and come back up all the way? Failure in this area could be an indication of a wreck. Open the trunk. Is there evidence of body work on the inside of the fenders? What is the general appearance of the trunk; Has someone been hauling gravel or bricks? What's the spare tire look like? Got a jack?

Get inside. How does it smell? A flood-damaged vehicle will smell moldy. Depending on how high the water got, you could be looking at some very unfunny rusting and electrical problems. If water entered the engine or transmission or rear end, failure is just a matter of time. Insurance companies auction off lots full of flood-damaged cars. Wholesalers buy them and ship them around the country to used car dealers who hide the damage the best they can and sell them. You can buy a flood-damaged car in Arizona that came from Georgia. Be suspicious if the car reeks of car freshener scents.

Does the interior generally give the appearance of a car that has been taken care of? This is a pretty good indication of how the owner takes care of the whole car.

Lift the hood. Is the engine reasonably clean or is it spotless? A spotless engine can mean the owner is a real automotive enthusiast or that he let the engine get super grubby and had it steam cleaned last week when he decided to sell it. Check down low on the side of the block or in hard to get to areas at the back of the engine to see if he missed a place with the steam. Then you'll know the story on steam cleaning the engine to go along with the "For Sale" sign. Check the battery water. Check the radiator. Low levels at either place means the owner doesn't keep tab on everything like he should. The water in the radiator won't be clear, but it shouldn't be rusty either.

Pull the oil filler cap and check the underside of it. Heavy deposits of carbon and sludge indicate abuse or a lot of miles. With a flashlight peer down in the oil filler opening as best you can. An engine with thousands of miles can be clean down in there, if the engine oil is changed on a regular basis along with the filter. Oil is normally black or very dark after it has been run, but it shouldn't be sludge-like on the dipstick. Oil should smell like oil—not gasoline. Water droplets on a dipstick can mean a bad head gasket, cracked block or head or water from being submerged. Ask the owner if you may take the top off the air cleaner. The element should be reasonably clean—certainly not half plugged up. If the owner won't let you maybe there is a reason you should know about.

Ask the owner to please start the engine with the hood up. You watch the engine. A lot of movement could indicate a broken engine mount. The engine should start smoothly and idle smoothly. A lot of smoke coming out the tail pipe means worn rings. A ticking sound could indicate a stuck valve lifter. A knock indicates a bad bearing on the crankshaft. If the car is equipped with an automatic transmission, check the fluid level while the engine is idling and warming up. Smell of the end of the trans dip stick. A burned odor indicates problems—running low on trans lube will burn band and clutch material.

In good lighting, check the body panels of a used car for alignment and smooth contours. Ripply surfaces may indicate major repair work and future trouble. This car looks O.K. Minor dings and scratches give you ammunition for serious price haggling but otherwise are not serious.

Check end of door and hinge area for evidence of repainting. Lift up rubber strips to check original color. Check the operation and condition of everything—doors, windows, controls, door locks, the whole thing.

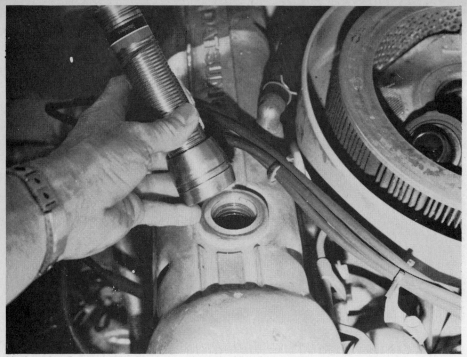
Underside of the oil filler cap plus whatever you can see down inside will give you some clues about accumulation of sludge and the general condition of the engine.

A new or recent air cleaner element speaks well for the previous owner's care of the car, *provided* all the other checkpoints show you the same thing. Be wary of cosmetic maintenance performed just before offering the vehicle for sale.

Take your time when checking all of this. Ask some questions. Did the owner buy it new? How many people now drive the car? Ever had any work done on the engine? Does he service it himself? Does he burn regular or premium gas? How many miles ago was the front end aligned? When was the last brake job done? What kind of gas mileage does the car get?

You're doing a couple of things with all of these questions. You are waiting for the engine to warm up so you can drive the car. You might be getting some solid information or you might be getting an indication the owner is lying through his teeth about everything. A front end alignment last month and a brake job last week on a car that has never had any engine work done in the last 98,000 miles and gets 23 miles per gallon despite the fact it is equipped with a 400 inch V-8 with an automatic is just a little hard to believe.

If you are still interested, go for a ride. Turn the radio on to see if it works, then turn it off. Run with the windows up for awhile, run with one down for awhile. Listen—for anything out of the ordinary. Does the transmission shift smoothly? Does the car run the way you want it to run? When you run it up to speed—say 50 MPH—back off the throttle and look in the mirror. Is there evidence of bluish smoke? Does the car lurch and wallow or does it feel firm and responsive. Does it pull to one side when going down the street or highway? Does it pull to one side when you apply the brakes? Is there a grinding noise when the brakes are applied? Does the brake pedal feel firm and is it near the top of the travel? On a stick shift car does the clutch seem to grab or slip or make noise? It should be smooth on engagement.

You should check everything: Wipers, power windows, air conditioning, heater, automatic door locks—everything.

When you park it, set the brake and let it idle. Turn on the headlights and go to the front. Both headlights working? Will they work on dim and bright? Pop the hood again. Remove the oil filler cap. Does smoke pour from it now that the engine is warm? The engine sound should not be much different now than it was when it was first fired up. Maybe smoother, maybe a little quieter—but not noisier.

If you want the car, make the owner an offer. Maybe he'll haggle, maybe he won't. Get the title to the car—not just the registration card—when you hand over the money. If the current owner is paying off the car through any kind of loan and you are at all unsure about how the transaction should be handled, go with the owner to the officer of a bank—preferably your bank—and you'll be briefed on what to sign and where to make it all legal.

When buying a used car from a lot, multiply everything we said about buying a new car from a dealer by about 200% and start shopping. Be ready for any story on anything having to do with the car—the way it looks runs, smells or where it came from. Turning back speedometers may be illegal, but it is still done and so are a lot of other things that are not illegal but will go a long way towards hiding the age or condition of a used car.

Used car lots normally spend $30 to $60 on a car before putting it on the lot for sale. Employees or "detailing" shops wash, wax, touch up paint, steam clean engines, scrub interiors and hide noises. They know how to give a car the appearance of a one owner (little old lady), low mileage, well cared for cream puff. These guys do a pretty good job, so take your time. If you can pull off the little inspection and driving drill we touched on earlier, you just might find some very good transportation.

USED CAR CHECK LIST

1. Check all body panels for any repairs—alignment, shape, overspray.
2. Check lower body panels for rust or attempts to hide rust.
3. Check trunk appearance.
4. Does car have spare tire and jack? Check condition of tire.
5. Check headlights on bright and dim, parking lights, turn signals, license light, tail lights, stop lights, back-up lights, interior lights.
6. Check doors for fit and all window operations.
7. Check interior for appearance—stains, moldy odor or attempts to hide same.
8. Operate all accessories—power windows, seats, radio, tape deck, air conditioning, defroster, heater, washer and wiper, horn.
9. Check front frame (under hood, under car) and suspension members for

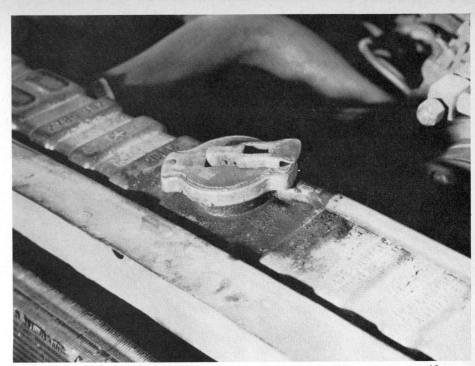

This *could* be water dribbling down on the top of the radiator from a leaky cap. If so, the repair cost is the price of a new cap—possibly two bucks. It *actually is* a crack where the filler neck joins the radiator top tank. Needs fixing by a radiator shop.

New battery and old corrosion raise questions about maintenance given by previous owner.

evidence of heating, welding, hammering, painting.

10. Check all tires for wear and size variation.
11. Check all fluid levels under hood—water in oil, etc.
12. Check oil-filler cap for sludge.
13. Check for smoke from oil-filler cap when engine is idling after warm up.
14. Check for evidence of recent steam cleaning or detailing.
15. Remove air filter element and check for condition.
16. Does car idle smooth, sound quiet?
17. Does automatic transmission make all shifts smoothly, evenly?
18. On stick shift, does clutch grab or slip?
19. Running gear quiet and smooth?
20. Car pull to one side at any speed or when braking?
21. Does car bottom on dips or bumps?
22. Smoke from exhaust on deceleration?
23. Check for evidence of leaks under engine, transmission and rear axle.
24. Exhaust quiet and leak free?
25. Owner or salesman nervous, anxious or just full of big stories?

SHOPPING FOR AN ECONOMY CAR LOAN

One of the best ways to save money when buying a car is to not borrow any money. There was a time when nearly everyone paid cash for their car but apparently that time has passed forever. The current thinking of a dollar down and a dollar a week may get you into the latest equipment, but it also can get you into debt by having you pay far more for the car than you intended.

Car loans carry a heavy penalty called interest. As this is being written new car loans are costing a lot of buyers as much as 12%. Used car loans are going for about 17%. You can pay even more. You can pay less—that's what this is about.

If you can't pay cash for a car, consider taking the money out of a savings account. Talk it over with your banker or savings and loan officer before you find the car you want. Without pressure, without a sales pitch decide where the money is coming from before being forced to make a hasty decision in a dealership. For instance, your banker might suggest that you use your savings account for collateral on the car. They keep your passbook; you

CHECK THE BLUE BOOK

Bankers, loan companies, car dealers and private individuals use a guide to the value of used cars. One is called the Kelley Blue Book. It is published six times a year and has been around for more than 40 years. You can subscribe or buy individual copies from Kelley Blue Book, P. O. Box 7127, Long Beach, CA 90807.

Whenever you are shopping for a used car and have found one that interests you, check the Blue Book. Even if you are buying from a private individual.

Consider it as a guide—not gospel—because prices are influenced by demand and market conditions as well as what the book says. However, the published book values probably have as much influence on the market worth of a car as any other factor.

A bank officer or loan company will usually be glad to show you a copy and will sometimes tell you values over the phone. Most used-car salesmen will let you look at a copy even though it's their secret weapon. They don't do it gleefully but they know it will turn you off if they don't and they know the next guy will let you see the book.

Kelley publishes wholesale and retail values of all cars based on clean units, fully reconditioned with acceptable mileage. There are mileage adjustment schedules showing how much to add or take from the value of the car when mileage is very low or quite high. Similiar schedules are in the book for optional equipment.

Suggested loan values are also given. A word needs to be said about this because if you are borrowing money to buy a used car you could get confused. Let's say you spot a car you want to buy. It is super clean. The owner is asking $100 over the Blue Book retail value. You agree it is a fair price since the car is in great condition. The Blue Book indicates a suggested loan value of 80% for that year car. That means a banker or loan company will normally loan you 80% of the value of the car. In the banker's eyes the value of the car is the wholesale figure—not retail and certainly not the higher-than-Blue Book-price you are willing to pay for the car.

The Blue Book carries a lot of information extremely useful to anyone buying or selling a used car. You'll hear the term mentioned plenty when you are around the used car market.

have a clear title to the car. The money in the account keeps right on earning interest. If you wish, you can continue adding to the account—just like car payments—but with the added benefit of getting interest on the payments. Naturally, you cannot make a withdrawal from the account until the loan is repaid but the interest you earn on your savings helps balance the interest you are paying on the loan.

If you don't have a passbook acount or the prospect of having your savings frozen doesn't appeal to you, start shopping for a loan. There are a lot of places to borrow money for a new car—maybe there are some you didn't think of. Credit unions, banks, savings and loans, insurance companies and car dealers themselves all have money available to the car buyer—but for a price.

If you belong to a credit union, check there first when financing a car. By Federal law, credit unions cannot issue auto loans with Annual Percentage Rates of more than 12%. Currently most credit unions are offering car loans of about 9%, but you might find something as low as 7 1/2%.

Probably the next best place to get money for a new car is a bank. Start with the bank you normally use, but it can pay to shop around. The current Annual Percentage Rate for a new car loan issued by a bank is 10 to 11%. What you are charged for a loan will depend on your credit rating, the state of the economy at the time you want the loan and competition among banks in the area. Do some shopping.

Anyone in the business of making loans such as banks and loan companies is required by the Truth in Lending Act to give you the true annual interest rate. This is called the Annual Percentage Rate.

This is what you shop for—the Annual Percentage Rate. Remember Federal law says the lender has to give that rate to you—orally and in writing. Don't go for any other wording. There are a lot of different ways of stating interest rates—but by comparing Annual Percentage Rates you can find out who has the best game in town. Do this no matter where you shop for a loan.

Check with your car insurance company for a loan. Surprised at that one? Most people never consider borrowing money from an insurance company—but it can be done. Annual Percentage Rates are often higher than you would find at a bank. But the insurance company might loan you money when the bank won't. A call to your insurance agent will give you the answer in a hurry.

If you have a life insurance policy with cash value you can probably borrow from it for most any purpose. Annual Percentage Rates on life-insurance loans are rock bottom—4 to 5%. Like the very attractive deal of borrowing against a passbook account, there's a catch in borrowing against the cash value of your policy. Thus the amount of the policy payable to your beneficiaries at time of death is reduced. Talk this over with your insurance man—he might show you how the interest buildup on your policy can pay back the loan over a long period of time.

The last place I would borrow money for the purchase of a car is a loan company or car dealer. Loan companies typically deal with poor credit risks and charge more for their money. There are loan companies charging the maximum Annual Percentage Rate allowed by state law. If you get a loan from a dealer he gets the money for you—from a bank, a loan company or one of the big auto-company finance arms such as General Motors Acceptance Corp. (GMAC) or Ford Motor Credit Co. On top of what the dealer gets the money for, he tacks on his own loan fee, finders charge or whatever he cares to call it.

It is sometimes difficult to get the prospective lender to state what the Annual Percentage Rate is. They want to dazzle you with low monthly payments and figures like 7% (of what?). Long term loans with the magic of low payments can cost you a bundle in financing charges. The finance charge of a two-year loan is often close to double that of a one year loan in terms of dollars going out of your pocket. For the standard three-year loan the amount you pay is often near triple that of the one year loan.

When borrowing, get the Annual Percentage Rate of the loan in writing and finance just as little for just as short a time as possible.

HOW MUCH DOES A CAR REALLY COST?

In mid-'74 the Department of Transportation ran a small study on the actual cost of owning three different-size cars. In any projection of this type some assumptions must be made. The following is assumed in the DOT study:

Each car will be bought new, driven 100,000 miles in 10 years and

---

Costs for 75 and later cars will be a lot more bucks than are shown below. The chart clearly illustrates it costs dollars to own and drive a car.

COST PER YEAR OVER TEN YEARS

CAR	YEAR 1 14,500 miles	YEAR 2 13,000 miles	YEAR 3 11,500 miles	YEAR 4 10,000 miles	YEAR 5 9,900 miles	YEAR 6 9,900 miles	YEAR 7 9,500 miles	YEAR 8 8,500 miles	YEAR 9 7,500 miles	YEAR 10 5,700 miles	TOTAL COST 100,000 miles
Standard four-door sedan Price: $4,251	$2,427.27 16.7¢ per mile	$1,807.03 13.9¢ per mile	$1,829.50 15.9¢ per mile	$1,720.36 17.2¢ per mile	$1,492.77 15.1¢ per mile	$1,492.56 15.1¢ per mile	$1,653.55 17.4¢ per mile	$1,251.60 14.7¢ per mile	$1,317.44 17.6¢ per mile	$900.28 15.8¢ per mile	$15,892.36 15.9¢ per mile
Compact two-door sedan Price: $2,910	$1,570.14 10.8¢ per mile	$1,432.19 11¢ per mile	$1,389.89 12.1¢ per mile	$1,375.04 13.8¢ per mile	$1,372.15 13.9¢ per mile	$1,345.03 13.6¢ per mile	$1,504.12 15.8¢ per mile	$1,149.58 13.5¢ per mile	$959.19 12.8¢ per mile	$782.20 13.7¢ per mile	$12,879.53 12.9¢ per mile
Subcompact two-door sedan Price: $2,410	$1,283.37 8.9¢ per mile	$1,165.10 9¢ per mile	$1,095.18 9.5¢ per mile	$1,270.07 12.7¢ per mile	$1,207.62 12.2¢ per mile	$1,210.60 12.2¢ per mile	$1,199.02 12.6¢ per mile	$1,130.03 13.3¢ per mile	$874.21 11.7¢ per mile	$717.90 12.6¢ per mile	$11,153.10 11.2¢ per mile

scrapped for $50.

Costs are figured in 1974 dollars. Gasoline is figured at 52¢ a gallon for all years.

Collision insurance is dropped at the end of the fifth year because the cars would no longer be worth enough to warrant it.

The DOT thinks of a *standard* sedan as something like a Chevy, Ford, Buick, Olds, etc. with a V-8, automatic transmission, power steering and brakes and air conditioning. A *compact* is defined as something on the order of an AMC Hornet, Plymouth Valiant, or Chevy Nova—all with six-cylinder engines, automatic transmissions and power steering. A *subcompact* is a car the size of a Chevy Vega, Ford Pinto, Datsun or Toyota. For purposes of the paper study all subcompacts were equipped with four-cylinder engines and standard transmissions.

Note that the cost per mile varies from year to year. There are several reasons for this. Depreciation is quite high in the first year for big cars. Normally cars are driven more when they are new. Then repair costs mount as the car gets older.

Costs and totals will be considerably more as inflation continues to erode the value of our money. With 1975 subcompacts selling in the $4,000 to $5,000 range, the chart is obviously understating the costs of owning and operating any car, regardless of size. But it gives you an idea of *relative* costs, at least.

NOW IT'S YOUR TURN

Unless you have a real love for bookkeeping, you probably won't go to all of the trouble the DOT did. But if you think you want to buy a new car to save some money—and you are very serious about the *saving money* part with no romance in the back of your head about buying a new car, then get a pencil and paper and be prepared for a little homework. What you learn could be an eye opener.

Start with the car you have. Find out what it's costing you. Don't fudge. You may be cringing before all of this is over, but maybe all of this will teach you to be a little more careful with your money.

Run a gas mileage check for a couple of weeks and find out how many miles per gallon you are getting. Also keep a record of how much that gas is costing. Drive normally.

During that two week period find out what your car is worth. In other words, if you had to sell the car in less than 10 days, how much could you get for it. Check newspaper ads, used car lots. Call a banker or used car dealer and ask if they would please quote you "high and low blue book" on your car. That's a standard pricing guide and it will get you close. If the car is just average—with a nick or ding here, value your car at the midway point between high and low blue book. If it is a real cream puff with low mileage, the value goes up. If your dog has eaten a hole in the back seat and all of the tires are bald, lots of luck.

If your car needs repairs, find out how much this will cost. If you don't know if you need repairs, take your car to a reputable garage. Tell them you are planning to keep the car for another two years and want troublefree transportation. Ask what needs to be done and how much it will cost. The more estimates you have the more you will know about your car—and the garages in your area.

The last time you bought a license plate for the car, how much did it cost, how much will it be next year? Look through your records or call the state agency and find out. While digging through your paper work, find out what you are paying for insurance. Now is a good time to call your insurance agent. Tell him what you are doing. Can he write you a new policy and save you money. If your driving record has improved since you last bought insurance—that can lower the cost.

While talking to your insurance agent tell him you are considering buying a Custom Plushmobile and ask what the insurance will be on this barge. Compile a "What's it gonna cost me?" list for the new car you might buy. Check gas mileage figures in a magazine road test—they're pretty accurate. If you use EPA figures, knock off about 20% to be realistic even if you are a reasonably conservative driver. Using that figure and your estimate of how many miles you'll drive in the coming year, you can come pretty close on what gas for the new buggy will cost you.

What will it cost you to buy the new car? How much interest in dollars, not per cent, will you shell out in the next 12 months if you buy a new car?

If you have car payments now, put those down as cost of your present car. A quick check with your banker will tell you what payments will be on a new car—once you tell him roughly how much the car will cost and how much you intend to put down on it. How much will the sales tax be on the new car? The car salesman can tell you that. He can also tell you what the license plates will cost for the first year.

Find out as best you can what your current car will cost you for operation next year including any foreseeable repairs, and do the same thing for the new car you propose to buy. The hidden costs of a new car are the ones that kill the bank account—interest on the note, sales tax, license plates, insurance—and we haven't even mentioned depreciation. Depending on how long you plan to keep the proposed new car, you simply might not want to consider depreciation. If you won't keep it more than three years or so—it's important. If you keep it 15 years, it's not so important.

Obviously, the more care and time you devote to this little financial report to yourself the more of a handle you will have on what it is costing you to own and drive a car.

Then you can make an intelligent decision balancing all the factors of additional costs and extra savings between the two propositions. You'll have a pretty good idea of what you're getting into either way—whether you keep the present car or buy a new one.

Probably 95% of car buying decisions are made strictly on emotion. If dollars are important to you, count the dollars and skip the emotional part.

7 SELL IT YOURSELF

The big advantage to trading an old car in on a new one is there is less hassle on your part. You drive the old hack in, agree on a trade-in value, buy the new and you drive out in it. This can be very simple—and very costly. The dealer intends to make money off every transaction. When you trade in a car, the dealer is essentially buying it from you. He gives you less than it is worth. He sells you a new car. He makes money. Then he sells your old car for more than he paid for it. He makes money. How are you doing?

Don't get down on car dealers. But realize and accept the fact they are businessmen. To remain in business, they must make a profit. There is nothing wrong with that. They may be in the business of selling cars, but they are also in the business of selling you a service too. This is what they do when they accept your clunker on a trade-in.

Look at your car with the eyes of a buyer—not a seller. Three tires are bald. The air conditioning doesn't work, one headlight is out, the paint is faded, a chrome strip is missing and you know the car is long overdue for a complete brake job and front-end alignment. O.K., now that you see the car in a different light, what do you do about it? Sell it like it is for a low price, do the work yourself and sell it for more money, or get rid of the eyesore. If the latter is your choice, then the dealer is doing you a service by taking it off your hands.

Let's say you decide to put a little money in the old heap and spend a couple of weekends fixing it up; then you plan to sell it yourself. It takes two weekends of steady work to get it in shape and a full month to sell it. Was it worth it? Only you can answer that. If you are busy on weekends, or hate working on cars, then this is probably the wrong route to take. A trade could be an easy way out—you're trading dollars for time and hassle. That can be a bargain.

SELLING IS SAVING TOO

Depending on what the dealer offers you for the old hack and the price of the new car with and without trade-in, you might come out several hundred dollars ahead by selling your own car. If you are undecided as to trading or selling before going to the showroom, do this: Find out what your car is selling for in the market. What will the car bring once it gets on a dealer's used car lot? Check blue book, used car lots, newspaper ads locally

You can trade your old car on another one, or just bring in your old barge and take what they offer. Used car dealer's motto is *Buy Low, Sell High.* If he buys low, you sell low.

and in nearby large cities. This is where time and knowledge of what your old car is worth pays off.

Never balk at selling your own car. Wax it, clean it up, put a card in the window with the sales price and your phone number. Tell people you have a car for sale. Shopping centers, supermarkets and large factories have bulletin boards for free advertising. Place neat, clearly worded 3 x 5 cards on any billboards you can. If this fails to produce the results you want, run some ads in the local papers. They're inexpensive.

There are two tactics to take on price. Set the price high, knowing full well you will haggle and the price will come down. If you don't have the patience or stomach for that, set the price at what you will take. Then when the haggling and lower offers start, you'll have to make it clear in a polite way that the price you have set is fair, it is firm—you know the quality of your merchandise. There will be lookers, tire kickers, bargain hunters, people wanting to trade shotguns and furniture for the car. There will be a buyer.

When a prospective buyer wants to go for a drive—go with him. It is not enough that he park his hack in your drive, leave you the keys to it and wheel off in your machine. Maybe he just swiped the thing he drove up in. Stick with the prospective customer at all times he is behind the wheel of your car or has the keys. Be polite, be firm, and be there. After selling a score or so of cars out of my various driveways over the years, I strongly suggest that when you come down to the sale, take only cash or cashier's check and nothing in between. If this seems hardnosed and offends someone, you're sorry. There are enough tricks and stories around to fill a book twice this size. It all adds up to the same thing—you lose. The only question is—how much. Try this one:

You are selling your car for $3500. The price is fair. You are firm. You have advertised the car and haven't snagged a buyer yet. It is Saturday afternoon. A nice looking young couple with a baby comes by to look at the car. You go for a ride with them. They like the car. They are polite and very interested. They'll call and let you know if they want to buy it. They leave.

Two hours later they call back. They want that car. It's more than they wanted to pay, but the car is nice and they are willing to pay the price. The young couple fully understands that you want cash or a cashier's check—but it is Saturday afternoon.

Can they write you a check for $800 as a deposit on the car until they can get to the bank Monday morning and get the cashier's check? Sure. They assure you they don't intend to pick up the car when they deliver the $800. They just want to make sure you don't sell it before they get to the bank. Sure. They deliver the $800 check. Everyone is all smiles.

Monday morning, the young lady arrives on the doorstep. A friend has brought her over to pick up the car. You're at work. Your wife is handling the transaction for you. The buyer has $2700 in cash. That plus the $800 makes $3500—full price for the car. Your wife hands over title and registration papers on the car and the deal is closed. The cash is real, the check turns out to be hotter than a fifty cent duck in a two dollar oven.

You were "had" for $800. The way to handle this one is to accept the $800 check, but make it very clear it is being used as a deposit only and you will hand it back to them when they come to pick up the car and deliver you the *full purchase price* in cashier's check or cash. The stories are endless—so is the grief.

SWAPPING

Trading one car for another is virtually a way of life for some people and for others it is totally foreign. There are those who would rather trade one item for another than accept cash and make a profit. They might not admit that, but it's true.

The pursuit, the haggling, the wheeling and dealing, the offers and counter offers culminate in "a deal." When you offer your car for sale to the general public you are placing your phone number and merchandise before people who don't want your car or even need it but will offer you something for it. Coon dogs, firewood, trailers, tents, firearms, used TV sets and other cars are traded daily for cars. This is a time-honored way of doing business and there is something to be said for it. Wealth need not be synonymous with cash or bank balance.

Some extremely good deals can be had by trading merchandise—good deals for both sides. The same goes for bad deals.

Recently I traded a compact for a late model full-size station wagon. I didn't need the compact and the people with the station wagon did. I needed a station wagon.

My compact was in good shape—so was their station wagon. The only strike against the station wagon was an expensive sound coming from the rear axle. I tracked the sound to a bad axle bearing. The bill to fix it came to $27. I'll never know for sure, but I suspect the previous owners had heard the sound, figured the repair would cost them several hundred dollars and decided to get rid of it in a hurry. I'm glad they did. For me it was a good trade.

Every deal is different. Know the value and condition of your merchandise and take some steps to learn the value and condition of the merchandise being traded. Good luck.

A little time and bargaining may make this sign in the window of your car worth an extra $300 or $500.

8 A SHORT COURSE IN MECHANICS

Only a few simple ideas are needed to understand the basics of engines. That's what this chapter is about. When you have read this part, you will understand the mechanical alterations and adjustments you can make to your car to improve economy and how to tune it up. You may not be able to actually do the mechanical things yet, but you will understand them. Chapter 9 shows you how to do a tune-up.

To perform mechanical work on an engine you need a shop manual which describes your exact model and has photos with labels to tell you where the parts and adjustments are. Shop manuals are nearly always available from the dealer who sells your brand although he may have to order it for you or supply an address for you to order it. If you can't get the factory manual for your car, check book stores and auto parts stores for manuals from independent publishers covering your brand and model.

The second purpose of this chapter is to make it easier for you to read and understand a shop manual. Most manuals are written with the assumption the reader is already a pretty fair mechanic and knows what all the words mean. If you are already comfortable reading a shop manual, you can skip this chapter.

THE PISTONS GO UP AND DOWN

That's pretty basic and true for more

Cutaway photo supplied by Chevrolet Public Relations shows you most of the engine parts discussed in this chapter: pistons in the cylinders, valves with valve springs, and the crankshaft down in the bottom of the engine. You don't have to understand about all of this engine, just a few basic things.

Side view shows ignition distributor behind air cleaner. Wires from distributor lead to spark plugs. Air intake is through that oval-shaped tube on the side of the air cleaner.

than 99% of automobile engines. The sprinkling of exceptions today are the rotary engines used in Mazda and rumored in other brands.

Pistons are round and fit in a hole called the cylinder or bore of the engine. They are connected with metal rods, called connecting rods or "con" rods, to the crankshaft of the engine so when the pistons move up and down the crankshaft goes 'round and 'round.

Basically, the engine is an air pump. If you used something like an electric motor or your strong right arm to rotate the crankshaft, and if you didn't feed it any gasoline, the engine would just pump air in one side and out the other.

There is more than one piston so passages are arranged to feed air to all of them and collect the air they pump out. These are called the intake manifold and the exhaust manifold. The exhaust manifold for a four-cylinder engine collects the exhaust from four cylinder and combines it to feed into a single exhaust pipe. A V-8 engine usually dumps four cylinders on the left side into a manifold which connects to an exhaust pipe and duplicates the arrangement on the right side—with another manifold and pipe.

The piston moving up and down in the cylinder works as an air pump in a way similar to a bicycle tire pump—it draws some air in and then pushes it out again. To do this, it needs a few more mechanical parts. It needs a head which is a metal casting that fits on the top of the engine and closes the top end of the cylinders. Thus a space is created between the top of the piston and the inner surface of the head. This space gets smaller and larger as the piston moves up and down in the cylinder. The space above the piston is usually called the combustion chamber because that's where the gasoline will burn when we feed some to the engine.

To work properly, the piston needs a good air-tight seal between the sides of the piston and the inside of the cylinder. This is accomplished by metal piston rings. They fit in grooves near the top of the piston and press outward against the cylinder wall to make a good seal.

Also, the engine needs some valves which are normally mounted in the head. One valve lets air in, the other lets it out. These are the intake and exhaust valves.

When we want air to go into the combustion chamber, we open the intake valve and *at the same time* move the piston downward in the cylinder. This causes a "suction" which draws air in. When we want air to go out of the combustion chamber we wait until the piston is moving upward—toward the head—and open the exhaust valve. This allows the piston to force the air out by making the space above the piston smaller and smaller in volume.

It's obvious the valves have to be timed to open and close in synchronization with the movement of the piston. Precision timing of the valve operation is done by a camshaft which is geared to the crankshaft so they both rotate when the engine is running. The camshaft has some "bumps" which press on the ends of the valves and cause them to open at the correct time. When the camshaft rotates so it is no longer forcing the valve open, a valve spring causes it to close. In most engines, the camshaft doesn't actually contact the ends of the valves. There are intermediate parts to transmit the motion—pushrods and rocker arm assemblies which work like little teeter-totters.

Now if we take a close look at this contraption: The crankshaft and camshaft are rotating, the pistons are moving up and down in the cylinders, the valves are being opened and closed by the camshaft, air is being pumped in and out of the cylinders by movement of the pistons, and the whole thing is a bit ridiculous because you are standing there rotating the engine with human-power. We need a better trick.

The trick is to get the engine to do all those things under its own power and have some excess power left over to drive the car along the road. Feed it some gasoline.

In a real-world engine, when the piston moves downward to draw in some air it also draws in some gasoline mixed with the air. The mixing is done in a carburetor mounted on the intake manifold and fed with gasoline by a fuel line from a fuel pump. The fuel pump takes gasoline from the tank and delivers it to the carburetor under pressure—similar to a city water system which delivers water to your house, pressurized so when you open a faucet the water comes out.

Because of the fuel pump, a supply of gasoline is available to the carburetor. It takes as much as it needs and stores it in a cavity called the float bowl. This is done automatically by a float which rests on the surface of the gasoline in the float bowl. When the fuel level drops the float moves downward and opens up a valve to allow more gasoline to flow in. When the proper level has been restored, the float has moved back up and automatically closes the valve. This valve is called a float valve or a needle valve. The great operating principle of the float bowl will be perfectly clear if you take the lid off a toilet tank and look to see how it automatically fills itself and shuts off.

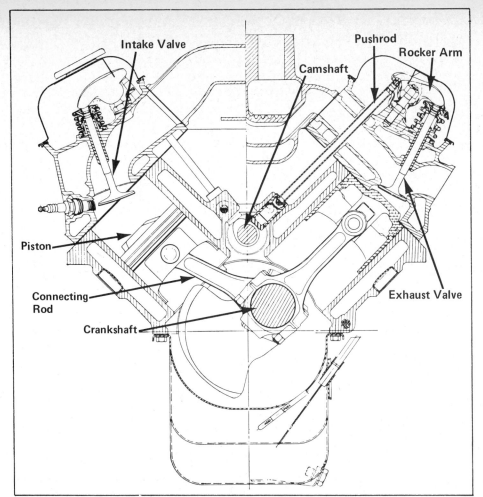

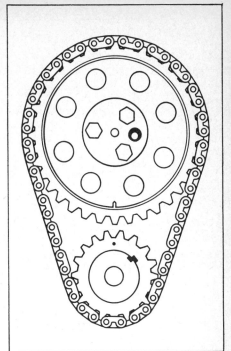

Camshaft which operates valves is driven by crankshaft using chain and two sprockets.

Cutaway view of a typical engine shows pistons connected to crankshaft so pistons can move up and down in cylinder bores. Valves on top of cylinders are opened and closed by rocker arms. Rockers are moved by pushrods which are operated by camshaft.

Holes like this, called *valve ports* lead to each cylinder of the engine. Larger opening lets in fuel-air mixture when valve is opened. Small port lets exhaust gases out.

If you look down through a big 4-barrel carburetor like this Holley, you should see the engine's tonsils or adenoids or something. Round tubes sticking out into each bore of this carburetor bring gasoline to be picked up by airstream through carburetor. The two "tanks" on each side are float bowls which contain a supply of gasoline. The float bowls are kept full by a fuel pump which draws from the gasoline tank which you fill at the service station too often.

Bottom side of carburetors such as this Rochester brand have round metal plates which open or close to control fuel-air mixture drawn in by the engine. These throttle plates are connected to your right foot.

Because of all that, a supply of gasoline is inside the carburetor float bowl and available to the engine. When the piston is moving downward *and the intake valve is open,* the piston draws air into the combustion chamber above the piston. The path for this air flow is: From the outside, through an air filter to remove dust and bugs, through a passage in the carburetor where some gasoline is added to the air flow, through the intake manifold and down the passage to the cylinder that is drawing in fuel and air, past the open intake valve and into the space above the piston. The engine is now drawing in a fuel-air mixture.

The air passage through the carburetor is called a venturi and as the air rushes through, it "sucks" gasoline from the float chamber. The gasoline is pulled through little holes called jets and sprays into the air stream where some of it vaporizes—it becomes a vapor instead of a liquid—and some of it travels along as little droplets in the air stream. The size of the jet—the size of the hole—has an effect on how much gasoline gets through the jet and into the air stream. Generally if you want to burn less gasoline you use a smaller jet size in the carburetor. But it's not ever that simple, as you will see.

The basic idea of carburetors is that air flow through the venturi causes a sort of vacuum or reduced pressure which draws gasoline into the air stream. When more air flows through the passage, it flows faster and creates more vacuum so more gasoline is drawn into the air stream. This is called the *venturi effect.*

The engine will reward you for giving it some gasoline by using the fuel-air mixture to drive itself and your car. It does this by burning the mixture in the combustion chamber above the piston. Burning raises the temperature of the air and increases the pressure in the combustion chamber. The increased pressure *forces* the piston to move downward which in turn forces the crankshaft to rotate.

To make the fuel-air mixture burn inside the engine, it has to be ignited; which is what the ignition system does. A spark plug is screwed into the head so the working end of the plug can ignite the mixture. When it is time to set it on fire, an electrical spark is caused to jump across a gap between two little metal pieces on the end of the spark plug. The spark raises the temperature in the vicinity of the spark gap and if the temperature gets high enough the fuel-air mixture will be ignited.

THE FOUR-STROKE CYCLE

Car engines operate by a four-stroke cycle. The word *cycle* as used here means a complete chain of events which repeats over and over.

The word stroke, in this case, means a movement of the piston from top to bottom *or* from bottom to top. The crankshaft rotates through half a revolution when the piston makes a downstroke and then another half-revolution when the piston makes the following upstroke. So, two strokes of the piston require one revolution of the crankshaft. Four strokes make two revolutions.

Two important strokes of the piston are *intake* and the *power* stroke. During intake, the piston moves from top to bottom drawing mixture in. During the power stroke, when the engine is extracting power from the burning mixture, the piston also moves from top to bottom because it is being forced downward by the combustion pressure above it.

Obviously, when the piston is at the bottom of the intake stroke it then has to move back up to the top again before it can start the power stroke. When moving back up, the volume of the combustion space above the piston is getting smaller and the fuel-air mixture in that space is being compressed by the piston. This is called the *compression* stroke.

The amount the mixture is compressed is stated by a number called *compression ratio*.

Compressing the mixture before ignition is more than just a happenstance. It is a very important part of the whole process of extracting power from gasoline. Development of the internal combustion engine demonstrated that higher compression ratios resulted in more efficient engines. The high-compression engine extracts more usable power from each gallon of gasoline and you get more miles per gallon.

Getting back to the four-stroke principle, we have so far described three—intake, compression, and the power stroke. It should be obvious that another upstroke of the piston must happen between the end of the power stroke when the piston is at the bottom of its

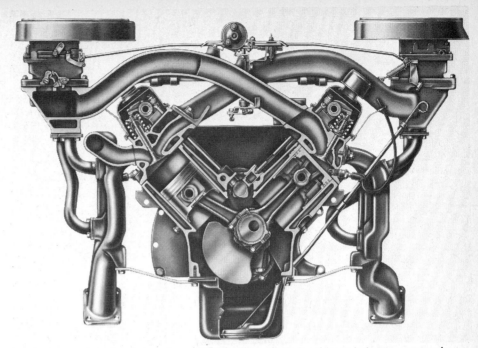

Turn on your imagination and you can "see" the fuel-air mixture being compressed as these pistons move upward in the cylinders.

travel and the beginning of the next intake stroke which starts with the piston at the top. This is the exhaust stroke.

At the end of the power stroke, the cylinder is filled with residue from combustion: Smoke, burned gases that are still hot, steam and miscellaneous pollutants. These can't be burned a second time so they have to be expelled to make room for the next charge of fresh fuel-air mixture to come into the cylinder.

The exhaust valve is opened and the piston moves from bottom to top, forcing the exhaust gases out of the cylinder and into the exhaust manifold.

Now the piston is at the top, ready to begin the cycle once more. The four events in the four-stroke cycle are: Intake, Compression, Power stroke, and Exhaust stroke. Each requires one stroke of the piston and as mentioned earlier the complete cycle requires two revolutions of the crankshaft.

Until recently, the trend in engine design and manufacture was to higher and higher compression ratios. Because these engines required better fuels, improved gasolines were developed with "anti-knock" additives.

Then came smog and the realization that the exhaust emissions of automobile engines was a major contributor to air pollution. There is more on this subject later. However, the need to reduce exhaust-emitted pollutants and the laws requiring such reduction have reversed the trends described above. Engine compression ratios have come back down so cars burn regular-grade rather than premium gasoline. A confusing array of emission controls have been installed on engines to reduce pollutants in the exhaust. None of these make gas mileage better and most make it significantly worse.

Finally it became obvious that gimmicks and gadgets on the engine were not the best answer to the problem and the increased fuel consumption was a major contributor to the energy crisis and the gasoline shortages of recent years.

An alternate approach is to build a more normal engine to burn gasoline and then use a special *catalytic* muffler to remove pollutants. Because the anti-knock additive—a compound of lead—makes a catalytic muffler stop doing its job, there is now lead-free gasoline available at service stations. Cars with catalytic mufflers should use this gasoline.

Time and events will tell if lead-free gas and catalytic mufflers are a good solution. There is a worthwhile gas mileage improvement because the engine can be built for more efficiency without the handicap of patchwork emission controls. However, these systems reportedly spew out raw sulfuric acid along with the engine exhaust gases. This can be harmful to people and vegetation and it may turn out that the cure was worse than the ailment.

IGNITION TIMING

Every multiple-cylinder engine has a cylinder firing order, determined by the design. A typical order for a four-cylinder engine is 1, 4, 3, 2, which means that cylinder number 1 delivers a power stroke, then number 4, then 3, then 2, then it starts over again with number 1. If the cylinders are all in a row, number 1 is usually the one nearest the front bumper. Engines with two cylinders on each side such as VW, or four cylinders on each side such as a V-8, will have a different cylinder numbering pattern. Whatever it is, the shop manual will show it by a diagram or drawing of the engine. Sometimes the cylinder numbers are marked on the engine itself.

Choosing the cylinder firing order is the engine designer's job and the choice is based on some fairly complicated technical reasons. Once chosen, the firing order is mechanically built-in and can not conveniently be changed. Knowing the firing order is part of your job if you are working on the enigne—which is why the manufacturer puts it in the manual.

The ignition system of an engine has two functions: Making a pulse of high-voltage electricity and then delivering that spark-pulse to the correct spark plug at the correct instant of time when the cylinder is ready to be fired. Let's look at how they are "timed" to be at the right cylinder at the right instant.

Electrical spark pulses are routed to the engine through a rotating electrical switch called a distributor. It's round, usually on the side of the engine or near the back, and it is gear-driven by the engine. A single ignition wire with thick insulation around it goes from the ignition coil to the top of the distributor and fits into a socket in a removable cover called the distributor cap. This wire feeds a series of spark pulses from the coil into the distributor. Inside the distributor is a rotating switch which *distributes* these electrical pulses to the correct cylinders according to the firing order of the engine.

For example, a distributor may deliver a spark pulse to cylinder number 1, then 4, then 3, then 2, then start over again.

Around the body of the distributor are ignition wires which conduct the electrical pulses to the spark plugs. There is one wire for each cylinder and at the distributor they are arranged in a circular pattern where they enter the body.

Inside the distributor is a rotating arm, called the rotor, which does the switching. It has a metal strip embedded in plastic and the metal strip is a conductor for the spark. When the rotor is pointing to the wire that leads to cylinder number 2, a spark is generated at that same instant and the spark goes to cylinder 2. Then the distributor rotor moves over to the wire serving the next cylinder in the firing order and the next spark goes to that cylinder.

If you take the cap off a distributor, you see the rotor and a circle of metal "posts" around the outside. These metal

The spark plug wires connect to a circle of metal terminals like this.

Inside the distributor, the spark voltage is distributed by a rotating switch with a moving rotor. The rotor serves one spark to each spark plug wire in sequence.

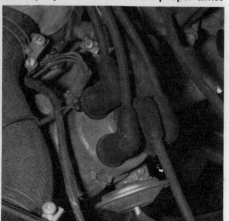

The center wire brings spark pulses into the distributor. The wires around the outside deliver spark pulses to each spark plug, in the proper order and at the proper time.

Underneath the rotor, inside the distributor, is a set of "points." This is an electrical switch—closed when the points are in contact with each other, open when they are separated. Points are opened and closed by a cam on the rotating distributor shaft. Opening the points makes the spark.

Electronic ignitions operate in a similar way except there are no mechanical points—it's all done by electronics.

"posts" or terminals connect to the ignition wires leading to the individual spark plugs.

The rotor can normally be removed simply by pulling it straight up and off the shaft. The shaft has a key and the rotor has a groove or keyway in it, so you can't get it back wrong.

Underneath the rotor are some electrical contact points which are opened and closed by a cam on the distributor rotor. These points make and time the ignition spark as follows: Current from the vehicle battery flows through the points when they are closed, then through the ignition coil which is mounted on the firewall or some convenient place. When the ignition points are knocked open by the distributor cam, the current flow into the ignition coil stops suddenly. This makes a very high voltage at the ignition coil for a very short length of time. We intend to use this high-voltage pulse to make a spark at one of the spark plugs.

But we have to get it to the correct plug. So, the high voltage pulse from the ignition coil is routed back over to the distributor by that fat wire that goes into the center of the cap. Then it is sent to the right spark plug by the action of the rotor.

The spark at a cylinder should arrive during the compression stroke, sometime before the piston arrives at top dead center (TDC). The reason is that the fuel/air mixture burns in a relatively short period of time, but not all at once. Because it takes a small amount of time for the burning to get underway and build up some useful pressure above the piston, we start it burning a little early. This is called *spark advance*. If you start it burning still earlier, you have more advance.

The mechanical arrangements described so far get ignition timing in the ballpark, but a method is provided to allow exact settings. The idea is simple—a moving part arrives at a non-moving part and gives it a whack. The moving part is the rotor shaft in the distributor with the ignition cam on it. The stationary part is the ignition point set which is mounted on the distributor housing. The housing does not rotate with the engine.

If you want to delay, or retard, ignition timing, you loosen a clamp screw which secures the distributor housing. Rotate the

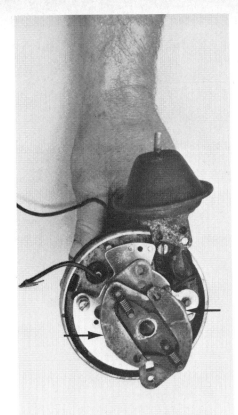

Weights inside the distributor move outward due to centrifugal force when the engine rotates faster, which changes ignition timing as needed. This is called automatic advance or centrifugal advance. Sometimes called mechanical advance. It makes the spark happen sooner when the engine goes faster, up to a point.

housing so the cam has to move farther to get to the place where it opens the points. This means you rotate the housing in the same direction the shaft rotates, to retard the spark.

To advance the spark, move it in the opposite direction.

Automatic Spark Advance—At idle speed, engines require some fixed amount of advance often called "static" advance, just to idle properly. Then as the engine is speeded up, more spark advance is needed up to about 3 or 4 thousand engine RPM. Above that the advance stays the same.

Inside a conventional distributor is an automatic spark advance mechanism with some flyweights and springs. The weights swing farther out from the rotor as they are whirled faster by the rotor. When they move outwards, they also change the ignition timing by rotating the ignition cam on the rotor shaft—in the correct direction. An automatic advancer that uses flyweights will put in more spark advance with increasing RPM, up to the maximum advance desired. This is often called *centrifugal advance* or *mechanical advance*.

Ignition timing is also sensitive to engine conditions which are indicated by the amount of vacuum in the intake system. Here is the idea of intake manifold vacuum. Between the carburetor and the engine is a throttle, usually mounted in the bottom part of the carburetor. This is a plate which is rotated by the gas pedal under your foot so it opens or closes the intake passage.

When the engine is running, the pistons

This enclosure contains a vacuum diaphragm which receives vacuum from the intake manifold. The diaphragm moves according to how much vacuum there is and that motion is used to put additional spark advance in the system during part throttle operation only. It's separate from mechanical advance illustrated above.

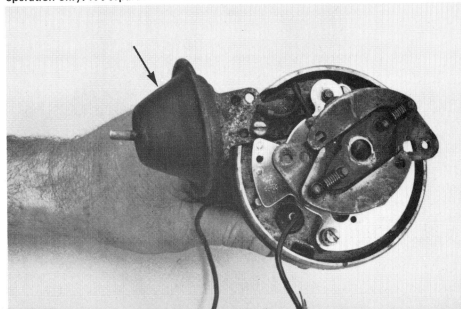

seek to draw in fuel-air mixture. If the throttle is closed, flow is reduced and a partial vacuum exists behind the throttle—that is on the engine side or in the intake manifold. If the engine runs faster, with the throttle closed, there is more vacuum. At wide open throttle, there is less vacuum. Therefore, manifold vacuum is a sort of *signal* which tells something about how fast the engine is running and how much the throttle is opened.

In general, manifold vacuum is an indication of load on the engine—high vacuum indicates light load such as idling or coasting downhill with the throttle closed. Low manifold vacuum means high load or open throttle.

Manifold vacuum is measured by "inches of mercury" which turns out to be a number such as 3 or 7. The higher the number of "inches," the more vacuum. Some distributors are built to control part throttle spark advance according to how much vacuum there is.

This is done by a diaphragm with one side exposed to outside air and the other side connected by a tube to manifold vacuum. When the manifold vacuum changes, the diaphragm moves and a mechanical linkage moves the mounting plate for the points. This is similar to you rotating the housing to set timing as part of an engine tune-up, but the vacuum diaphragm does it all the time the engine is running and vacuum is in the range of 4 to 12 inches.

How a carburetor works

This section and the one following on emissions have been excerpted from the H.P. book *Rochester Carburetors*.

All brands of modern carburetors have the same general features although some accomplish specific tasks in different ways. If you are unfamiliar with carburetors and their functions, this section will give you a good basic understanding. If your car does not use a Rochester carburetor, you should consult a shop manual or repair manual for your brand before tuning it.

Understanding how your carburetor works is the key to getting top performance from your engine-carburetor combination. And, it's also the quick way to get more performance with less work because you will know what is happening and be able to do your tuning in a meaningful way without wasting time on a lot of cut-and-try changes which could wreck your carburetor and/or leave it in a worse state of tune than when you started.

Carburetors are really very simple and similar devices, as you will see. Just as all four-cycle engines—from one-lung lawnmowers to Chrysler hemis—operate with the same principles of intake, compression, power and exhaust, all carburetors operate similarly when you compare them.

Regardless of how many venturis (barrels) your carburetor has, let's keep things really simple by starting out with a one-barrel carburetor. More complex carburetors are literally several one-barrel carburetors built into a single unit.

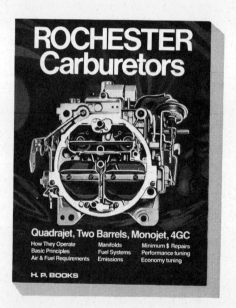

The carburetor mixes air and fuel in the correct proportions for engine operation. To satisfy the range of requirements of engines and drivers, several different systems must be employed: Inlet (float), idle, part-throttle, power, accelerator pump and choke.

Let's look at these basic systems one at a time and see what parts do each job.

FLOAT SYSTEM

The float (inlet) system consists of three major items:
1. Fuel bowl
2. Float
3. Inlet valve (needle and seat).

Fuel for the carburetor is stored in the fuel bowl. The fuel-inlet system must maintain the specified fuel level.

Movement of the needle in relation to its seat is controlled by the float which rises and falls with fuel level. As fuel level drops, the float drops, opening the needle valve to allow fuel to enter the bowl. When the fuel reaches a specified level, the float moves the needle to a position where it restricts the flow of fuel, admitting only enough fuel to replace that which is being used. Any slight change in the fuel level causes a corresponding movement of the float, opening or closing the fuel-inlet valve to restore or maintain the correct fuel level.

Some 1970 and later cars have a vent which connects the float bowl to a charcoal canister when the engine is turned off. This part of the emission-control system collects escaping fuel vapors. The same canister also collects fuel-tank vapors via plumbing from tank vents. Vapors collected in the charcoal canister are sucked back into the intake manifold through a system of valves when the engine is running.

Every carburetor has a published float setting established by the design engineers. The float setting is the float location which closes the inlet valve when the required fuel level in the bowl is reached. This is referred to as the "mechanical setting" as a measuring instrument is used to determine its position.

Inlet Valve—The rounded end of this valve rests against or is connected to the float-lever arm with a pull clip. The

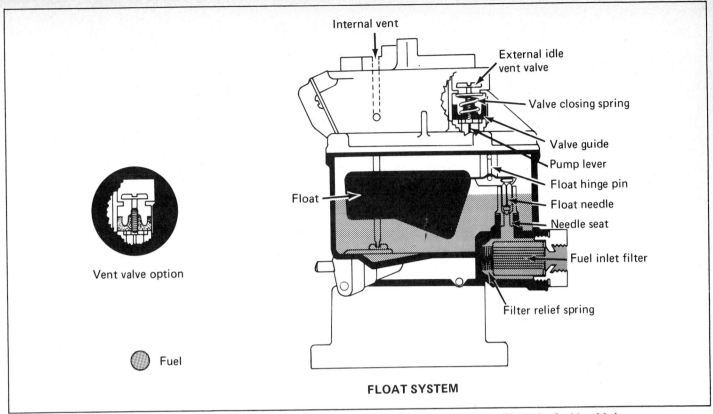

FLOAT SYSTEM

Basically, the float chamber of your carburetor keeps gasoline at a certain level by float action. When the fuel level is low, the float moves down, moves the float needle and allows fuel to flow in. When float rises to normal level, it pushes float needle against needle seat and closes off the entry. The water level in your favorite toilet tank is maintained in the same way.

tapered end of the inlet needle closes against an inlet-valve seat as the float rises.

The steel inlet valve needle in today's carburetors has a tapered seating surface tipped with Viton. Viton-tipped needles are extremely resistant to dirt and conform to the seat for good sealing with low closing forces.

A fuel filter or screen may be found in the carburetor body or fuel-bowl cover as part of the fuel-inlet system. The filtering device is placed between the fuel pump and the inlet valve to trap dirt which could cause inlet valve seating problems. This filter is installed with a pressure-relief spring behind. If the filter becomes clogged from excessive amounts of dirt, the relief spring lets the filter move off its seat. This ensures fuel flow to the carburetor, even if the filter is clogged.

MAIN SYSTEM

The main-metering system supplies air/fuel mixture to the engine for cruising speeds. During the time that engine speed or air flow is increasing to a point where the main-metering system begins to operate, fuel is fed by the idle and accelerator-pump systems, which are described separately. Under conditions of high load, when the engine must produce full power, added fuel comes from the power system, described in another section.

On its way to the cylinders, the air passes through the venturi. The venturi is a smooth-surfaced restriction in the path of the incoming air. It "necks down" the inrushing air column, then allows it to widen back to the throttle-bore diameter. Air is rushing in with a certain pressure. To get through the necked-down area (venturi), it must speed up, reducing the pressure inside the venturi.

Because the fuel bowl is maintained at near-atmospheric pressure by the vent system, fuel flows through the main jet and into the low-pressure or vacuum area in the venturi.

Pressure-drop (vacuum) at the venturi varies with engine speed and throttle position, increasing with engine RPM. Wide-open throttle and peak RPM give the highest flow and the highest pressure difference between the fuel bowl and a discharge nozzle in the venturi, thus the highest fuel flow into the engine.

The pressure drop in the venturi also depends on the size of the venturi itself. A small venturi provides a higher vacuum at any RPM and throttle opening.

Once main-system flow is started, fuel is metered (measured) through a main jet in the fuel bowl.

Many carburetors use a small *boost venturi* inside the larger main venturi. Boost venturis act exactly the same as the larger venturi, but supply a stronger vacuum to the discharge nozzle.

Boost venturis allow using a much shorter main venturi so the carburetor can be made short enough to fit under the hood of an automobile. Some carburetors use as many as two boost venturis to get adequate vacuum for main-system operation.

Main Jets—These metering orifices are used to control fuel flow into the metering system. They are rated in flow capacity and are removable for carburetor-calibration purposes. The final selection of the correct jet for the application will have to be done by testing because design and operational variations (climate, altitude and

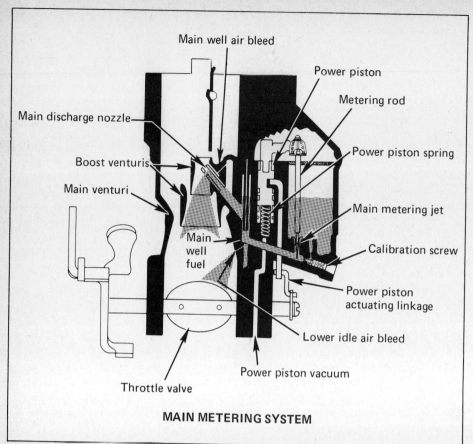

MAIN METERING SYSTEM

The reduced diameter main venturi causes a reduced pressure or vacuum near the Main discharge nozzle where fuel comes out into the air stream. The lower the pressure in the venturi, the more fuel is sprayed out by the main nozzle. Smaller venturis, identified as Boost Venturis help maintain low venturi pressure at low engine speeds.

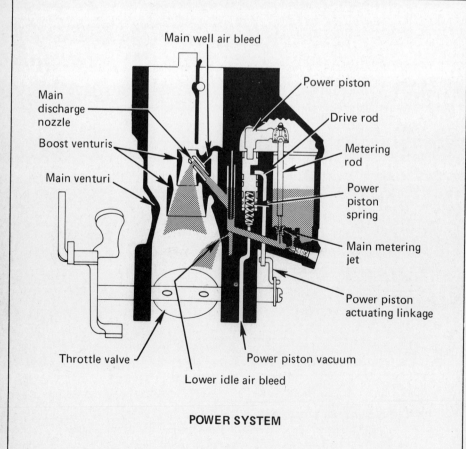

POWER SYSTEM

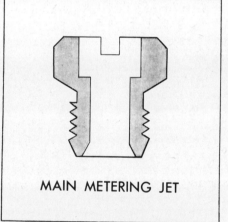

MAIN METERING JET

Carburetor jets control the amount of gasoline that can flow through them by the size and shape of the hole in the center of the jet. Jets are threaded so they can be changed if necessary by a carburetor specialist.

Power valves take different forms in different carburetors. In this model, additional fuel is fed to the engine by lifting up the metering rod which allows more gasoline to flow through the Main metering jet.

temperature) affect jet-size requirements.

Drilling out jets to change their size is never recommended because this always destroys the entry and exit features to a certain degree, and may introduce a swirl pattern, even if the drill is held in a pin vice and turned by hand. You cannot be sure of the flow characteristics of a jet that you modify by drilling.

POWER SYSTEM

When the engine is called on to produce power in excess of normal cruising requirements, the carburetor has to provide a richer mixture.

A vacuum passage in the carburetor applies manifold vacuum to a power-valve piston. At idle or normal-cruising-load conditions, manifold vacuum acting against a spring holds the valve closed. As high power demands load the engine, and manifold vacuum drops below a pre-set point, usually at about 7 inches of mercury (Hg), the power-valve spring overcomes manifold vacuum and starts to open the power valve. Most power systems progressively open to a full-flow position as manifold vacuum decreases.

The power valve can be considered as a switch which turns on extra fuel to change from an economical cruising air/fuel mixture to a power mixture. The power valve itself is merely a gate operated by manifold vacuum and spring pressure. It is designed to operate at a given load.

When engine power demands are reduced, increasing manifold vacuum acts on the power piston to overcome the spring tension, closing the power valve and shutting off the added fuel supply.

The power-valve opening point is another variable which the carburetor designer uses to arrive at the best compromise between economy, exhaust emissions, driveability and performance.

ACCELERATOR PUMP SYSTEM

The accelerator pump has three functions:

1. To make up for the fuel that condenses onto the manifold surfaces when the throttle is opened suddenly.

2. To make up for the lag in fuel delivery when the throttle is opened suddenly which allows more air to rush in without picking up enough fuel.

3. To act as a mechanical injection system to supply fuel before main system starts.

The accelerator pump operates when the pump-operating lever is actuated by throttle movement. As the throttle opens, the pump linkage operates a pump plunger. Pressure in the pump forces the pump-inlet ball or floating cup onto its seat so that fuel will not escape from the pump into the fuel bowl. Pressure also raises the discharge needle or ball off its seat so that fuel is discharged through a "shooter" into the venturi.

IDLE SYSTEM

Idling requires richer mixtures than part-throttle operation. Unless the idle mixture is richer, slow and irregular combustion will occur due to the high dilution of the charge by residual exhaust gases which exist at idle vacuums.

The idle system supplies fuel at idle and low speeds, and should keep the engine running even when accessory loads are applied to the engine. These include the alternator, air conditioning and power-steering pump. The idle system also has to keep the engine running against the load imposed by placing an automatic transmission in one of the operating ranges (low, drive, reverse).

When you stomp on the gas pedal, the pump plunger in the accelerator pump delivers a squirt of fuel into the carburetor air passage. By easing into the throttle more gently, you can reduce the "pump shot" of gasoline on most cars and save a little bit.

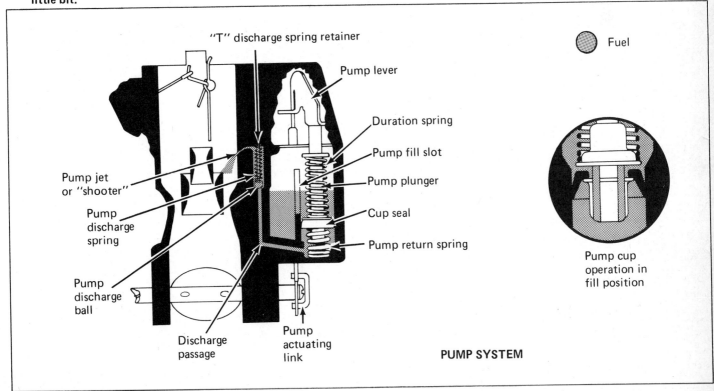

PUMP SYSTEM

At idle and low speeds, not enough air is drawn through the venturi to cause the main-metering system to operate. Intake-manifold vacuum is high because of the great restriction to the air flow by the nearly closed throttle valve. This high vacuum provides the pressure differential for idle-system operation.

When the throttle is closed—or nearly so—the reduced pressure between throttle and intake manifold draws air/fuel mixture through the curb-idle port in the bore *below* the throttle plate. When the throttle is just opening, manifold vacuum is still high and additional mixture is drawn through the off-idle port as it is uncovered by the opening throttle. The amount of flow through the idle system depends on the channel restriction, the size of the idle-discharge ports, and the idle-mixture needle setting.

As the throttle approaches wide open, there is low vacuum at the idle and off-idle ports and the idle system gradually ceases to deliver air/fuel mixture.

The idle ports are very small and are not intended to supply all of the fuel and air required by the engine even at a slow idle. The throttle is not closed completely at idle, or outside air comes around the throttle in some way. This air flow directly through the main bore of the carburetor is not enough to start up the main fuel system, so it is air alone, not air mixed with fuel. The air/fuel flow through the idle system must carry along enough fuel to make a proper idling ratio when added to the outside air flow past the throttle.

A throttle stop or adjustment which regulates the amount of air flowing past the throttle is used to set idling speed. The idle-mixture needle, which regulates flow through the idle system, is used to set air/fuel ratio.

In normal driving, flow swings quickly back and forth between idle and main operation as the vehicle is accelerated, slowed by closing the throttle, idled at stop, and then reaccelerated.

Idle-Speed Setting—Before emission-control requirements became important, idle setting was typically the slowest speed at which the engine would keep running smoothly. Emission requirements have made higher idle speeds necessary in many cases.

Cars with engines designed to pass emission requirements are typically set

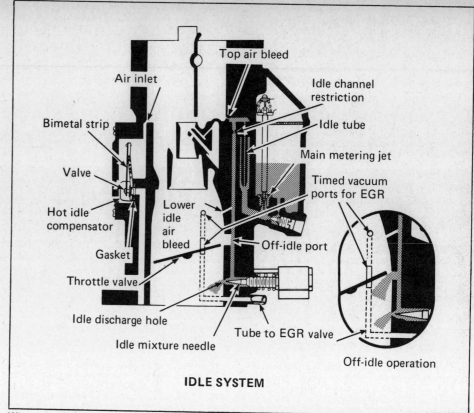

IDLE SYSTEM

When the throttle valve is closed or nearly so, fuel gets into the engine through a special set of tunnels and passages called the idle system. Turning the idle mixture needle makes the idle mixture more rich or more lean. The inset drawing shows that there is more than one discharge hole for idle mixture. The extra holes start dumping fuel into the carburetor as they are uncovered by the opening throttle plate.

for lean best idle at specified RPM and a subsequent reduction in idle speed by leaning the mixture still further—as stated on a label in the engine compartment.

Older non-emission-controlled cars and racing cars are typically idle-set for the desired idle RPM and best manifold vacuum. This is not a minimum-emission setting, however.

CHOKE SYSTEM

The choke system provides the richer mixture required to start and operate a cold engine. Cranking speeds for a cold engine are often around 50 to 75 RPM. These speeds are low compared to engine operating speeds, hence very little manifold vacuum is created to operate the idle system. A closed choke valve, which conforms closely to the inlet air horn, causes a vacuum below it so that fuel is pulled out of both the idle and main-metering systems during cranking. At times fuel is even pulled from air-bleed holes.

This surplus fuel creates an extremely rich mixture of approximately equal fuel and air. The super-rich mixture is needed because there is not much manifold vacuum to help vaporize the fuel and the manifold is cold, so most of the fuel puddles onto the manifold surfaces as it immediately recondenses. Liquid fuel cannot be evenly distributed to the cylinders and when it arrives there, it will not burn correctly. Only a small portion of the fuel ever reaches the cylinders as vapor during starting.

Once the engine starts, off-center mounting of the choke-plate shaft causes air flow to open the choke partially against torque of the bimetal spring so that the mixture is leaned out somewhat.

In the case of the automatic choke, there is a vacuum "break" diaphragm which pulls the choke valve to a pre-set opening once the engine starts. This opening provides a partial lean-out of the starting mixture. When the choke assumes this position, it is still providing a 20 to 50 percent richer-than-normal mixture during the warm-up period. This mixture is further leaned out towards a normal

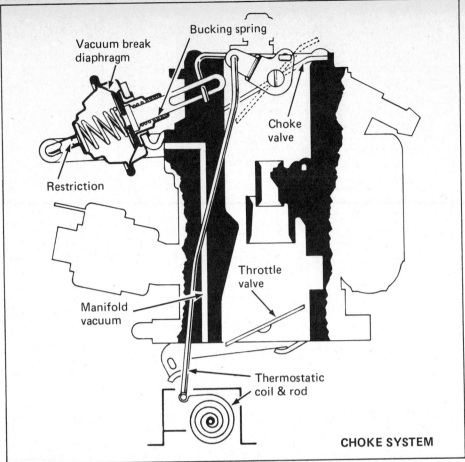

CHOKE SYSTEM

Starting a cold engine requires a rich mixture. The choke valve at the top of the carburetor is closed so there is high vacuum everywhere below it. Fuel gushes out of all openings and the mixture is very rich.

The choke valve is positioned automatically in modern cars by a thermostat which senses hot and cold. When the engine is cold, it closes the choke valve. When hot it opens the choke valve, automatically.

In smog-controlled automobiles a vacuum-break diaphragm is used to jerk the choke partly open before the engine is warmed up enough for the thermostat to do it. When the engine starts to run, manifold vacuum is fed to the vacuum-break diaphragm and that opens the choke partially whether the engine is ready or not. That's why your late-model car stalls about six times when you are trying to get it going in the morning.

operating mixture as the engine warms up and the bimetal choke thermostat weakens, allowing the choke to come off (open fully). Up through the '60s, chokes were calibrated to be off in 1.5 to 3.0 miles city driving. Today's chokes are for the most part effectively off at 0.8 to 1.5 miles as part of the current emission-reduction efforts.

Although it appears to be a simple valve atop the air horn, the choke system is often one of the most complex in the carburetor.

Fast Idle—During the warm-up period the engine has higher frictional forces to overcome. To prevent stalling, the engine must run at a higher idle speed than would be required if it were warm. A fast-idle screw on the throttle contacts a fast-idle cam linked to the choke which holds the throttle farther open than a normal idle. When the choke valve moves to its fully open position the fast-idle cam moves out of the way of the fast-idle screw so idle can return to its normal curb-idle setting.

SECONDARY THROTTLE OPERATION

Modern carburetors use small *primary* venturis for low engine speeds and then open up larger *secondary* venturis for high speeds when more air flow is needed. Operating the two venturis in sequence is called *staging*.

The late 1940's saw increased emphasis on vehicle performance. Because more air flow means more power, single carburetors became bigger and bigger—and so did the driveability problems.

Simply stated, the single-or two-barrel carburetor's metering range was too narrow to satisfy all requirements. The answer was obvious: Use a *staged* carburetor to stretch the metering range and get back to venturis small enough to get the main systems flowing at low RPM and provide good vaporization—while having the required capacity for high-RPM operation when needed.

Six-cylinder engines of medium displacement remained the economy workhorses, so no staged two-barrels were ever used on a U.S. car until the 1970 Pinto and the 1973 Vega, even though such systems had long been used on European and Japanese cars.

However, in the early 1950's, a whole host of four-barrel carburetors were introduced for V-8's. The primary side of these carburetors was smaller than it had been on the single-stagers, returning all of the benefits provided by small venturis. The primary carburetor barrels were used for the cruising loads and light accelerations encountered in normal traffic. Flexibility of operation and economy were regained.

Coupled to the primary carburetor barrels were two secondary carburetor barrels—designed to operate when maximum air flow was required for more power. Essentially, the metering range of the carburetor was doubled, combining good part-throttle operation with relatively unrestricted flow for maximum-power conditions.

Secondary carburetors are simply another carburetor in parallel with the first one. They always have their own main metering system. Some have their own idle system to one degree or another because this gives better distribution and idle or off-idle stability, thereby improving emissions by allowing the use of leaner idle settings.

Some of the carburetors have power systems in the secondary side.

The secondary side of a carburetor has no choke. Lockout devices prevent secondary throttle operation when the choke is on. There would be extreme engine stumbling due to lean mixtures if the secondary throttles were allowed to oper-

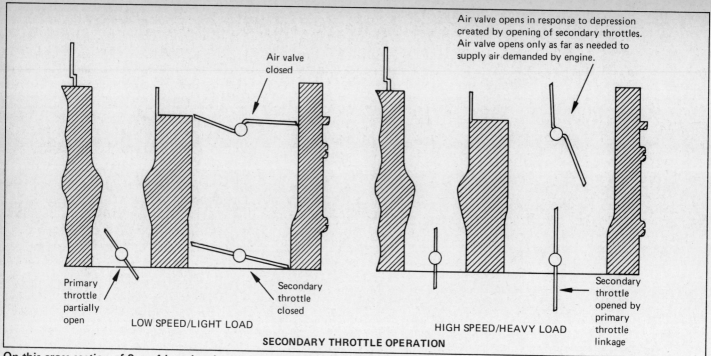

On this cross-section of 2- or 4-barrel carb, primary venturi is at left. Secondary venturi has a throttle plate at bottom and a special air valve at top. You can open the secondary throttle with your right foot, but the air valve protects you from folly. It won't open up and let air flow through the secondary until the engine really needs the extra fuel and air.

ate while the primary side choke was on.

Operation of the secondaries can be accomplished in several ways.

Mechanical—The secondary throttles are opened by a direct mechanical linkage from the primary throttles.

Diaphragm—Secondary throttles are opened by a vacuum-operated diaphragm. This adjsuts the amount of throttle opening to engine demand.

Mechanical with Air Valve—This is another way to adjust secondary airflow to engine demand. See the drawing above.

Mechanical with Velocity Valve—Mechanically different but functionally similiar to Air Valve control of the secondaries.

Use of diaphragm, Air Valve and Velocity Valve control methods has the advantage of opening the secondary throttles only when there is enough air flow to require it. At low RPM, a full-throttle demand from your right foot will only open the primary throttles.

OTHER FUNCTIONS

The carburetor also provides control for other important functions, especially those related to spark advance/retard and emission controls. Signals created by various pressure areas in the carburetor, especially as related to throttle position, are used to good advantage by automobile designers.

Vacuum Advance Ports—Some carburetors have a slot or holes drilled just above and alongside (not connected!) the off-idle discharge slot or holes. These feed vacuum to operate the distributor vacuum advance unit when the throttle is opened past idle. You will sometimes hear this carburetor-controlled distributor vacuum advance referred to as *timed spark*.

Canister Purging—Certain 1970 and all 1971 and later GM model cars have completely closed fuel tank venting to control evaporative emissions. The vent from the fuel tank leads into a vapor collection canister.

Because the fuel tank is not vented to atmosphere and the carburetor is vented only to the canister when the engine is stopped, fuel vapors are collected in the vapor canister. Purge ports for the canister are provided in the carburetor throttle body on certain models. The purge ports lead through passages to a common chamber in the throttle body to a purge tube hose connected to the vapor canister. The purge ports may consist of a constant-bleed purge and a separate timed canister purge, or a separate timed canister purge only.

The constant-bleed purge operates during idle, to purge the canister continuously when the engine is running. Along with the constant-bleed purge for the canister, a timed purge may also be used. The timed purge port is located in each bore adjacent to the off-idle discharge ports. The timed purge operates during off-idle, part-throttle and wide-open throttle. This provides a large purge capacity for the vapor canister and prevents over-rich mixtures from being added to the carburetor at any time.

The carburetor and emission controls

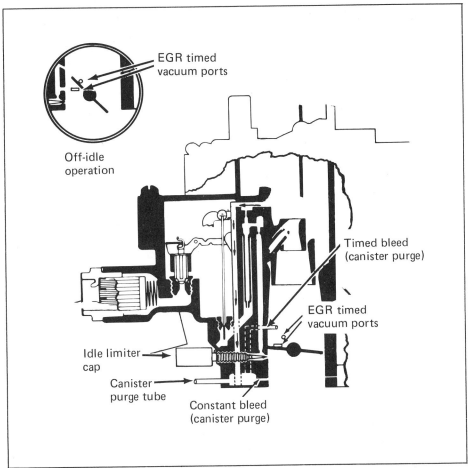

Recent cars have a charcoal canister which traps fuel vapors from tank and carburetor to reduce smog. When the engine is running, these vapors are drawn into the carburetor through a canister purge tube and the vapors are burned in the engine. This drawing also shows special ports in the side of the carburetor near the throttle plate. Vacuum from these ports is influenced by throttle position. The vacuum is used to control anti-pollution devices on the engine.

What is *smog*, anyhow?

It is a simple term for a complex happening. When unburned hydrocarbons (HC) and oxides of nitrogen (NO_x) combine in the atmosphere and are acted upon by sunlight, complicated chemical reactions occur to produce *photochemical smog*.

Smog causes nose, throat and eye irritations. And, like carbon monoxide (CO), is extremely harmful to animal and plant life (including trees). Smog also causes deterioration of some plastics, paint, and the rubber in tires, seals, weatherstripping and windshield-wiper blades.

There are three major types of vehicle emissions to consider:

1. Crankcase
2. Exhaust
3. Evaporative

In this section there are facts intended to acquaint you with the emission equipment and how it affects you and your car. Theoretically, perfect combustion of gasoline and oxygen would produce only carbon dioxide and water as emissions. Unfortunately, pressure and temperatures in the combustion chamber cause several undesirable combustion products. Two primary ones are hydrocarbons and carbon monoxide. They are nothing more than partially burned air/fuel mixtures. Partial combustion is caused by a shortage of oxygen in the mixture, combustion chamber shape, cam overlap and myriad other factors—all related.

Carbon monoxide forms whenever there is insufficient oxygen to complete the combustion process. Generally speaking, the richer the mixture, the higher the CO concentration. Even if the fuel/air mixture is chemically correct, CO cannot be reduced to zero because perfect mixing and cylinder-to-cylinder distribution is impossible to achieve.

Gasoline is composed of numerous and varied hydrogen and carbon compounds, hence the name *hydrocarbons*. Unburned hydrocarbons are just that—gasoline that did not get burned on its trip through the engine. There are several reasons why this can and does happen. Rich mixture is one. Fuel that does not get burned because of misfiring is another. So, either a lean or a rich mixture can cause an increase in HC emissions. Other things affect HC concentration.

A third product emitting from combustion chambers is oxides of nitrogen. These form when the normally inert nitrogen present in the air/fuel mixture combines with oxygen under high temperature and pressure.

To be able to pass emissions tests and also to make the car reasonably driveable, the carburetor is calibrated *too-lean* at the lower speeds and *over-rich* at the higher speeds and for heavy throttle maneuvers, such as passing and climbing grades. In some cases this is particularly bad for altitude operation.

Let's not misunderstand this seeming criticism of what has been done to date to combat motor-vehicle emissions. It is for us, the motorists, that billions of dollars have been spent on this task. The question is: Is progress being made in the *right* direction? Regulations are being met by the manufacturers, but does all of this *reduce* emissions?

In addition to the emission tests

Emission controls on this 1973 Chevrolet engine include air pump and distribution plumbing for AIR system, EGR (arrow), PCV, and ECS. If you want to know what all that stuff is, read about it in the accompanying text.

The PCV valve saves fuel because combustion products and some unburned fuel—called blow-by—leaks past the piston rings. If allowed to remain in the crankcase they do bad things. Positive Crankcase Ventilation starts with a breeze of fresh filtered air from the air filter into the crankcase. From there it is drawn through the PCV valve, taking the blow-by with it, into the intake manifold and finally into the combustion chamber where it gets a second chance to burn.

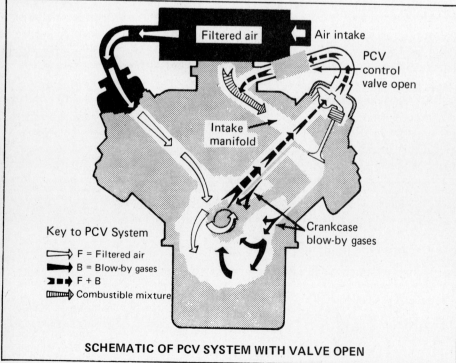

SCHEMATIC OF PCV SYSTEM WITH VALVE OPEN

required for exhaust-gas analysis, another test determines total hydrocarbons emitted from a non-operating vehicle.

A vehicle is placed in a sealed enclosure called a "bag." The evaporating hydrocarbon emissions are then measured as they accumulate in the enclosure.

VEHICLE EMISSION EQUIPMENT

This section will identify and show the purpose of the different emission-control devices introduced since 1959. These components will be illustrated and commented on so the novice and mechanic alike can understand what these mechanical units do.

POSITIVE CRANKCASE VENTILATION (PCV)

Crankcase emissions were the first target of the lawmakers and automotive engineers. They were first because approximately one third of all engine emissions come from this point. And anyone could look at the old road-draft vents—used on all cars and trucks through 1960—and see lots and lots of pollutants being spewed into the atmosphere. If you stand near an early car (pre-1965 in most of the United States) the smell of the escaping hydrocarbons and carbon monoxide is unmistakable.

The gasoline engine combustion process creates a highly corrosive gas. And, for every gallon of gasoline burned, more than a gallon of water is formed. Some unburned fuel and combustion products leak past the piston rings into the crankcase as *blow-by*. This blow-by must be removed before it condenses in the crankcase and reacts with the oil to form sludge. If this sludge is circulated with the oil it will cause corrosive and accelerated wear of pistons, rings, valves, bearings, etc. And, because blow-by carries a certain amount of unburned fuel, oil dilution also occurs if it is not removed.

The first emission-control device—required by California in 1961—was subsequently required nationwide. This was a crankcase ventilation system including a metering valve between the crankcase vent and the intake manifold.

By 1968 the Positive Crankcase Ventilation (PCV) system was the standard crankcase ventilation system on all U.S. cars. It removes engine crankcase vapors resulting from normal engine blow-by. These are removed by using manifold vacuum to draw fresh air through the crankcase, sucking the undesirable gases and unburned fuel into the manifold so they can be burned in the engine.

The PCV valve varies the flow through the system according to engine operation. The valve is spring-loaded. As manifold vacuum drops, the spring opens the valve so PCV flow increases with engine speed.

As shown in the accompanying diagram, when the PCV valve is open air flows through the air cleaner, through the crankcase where it picks up vapor, through the PCV valve, and into the intake manifold.

At idle, or whenever the throttle is closed, the PCV valve also closes except for a small fixed passage so an excessive amount of air is not drawn into the manifold where it would lean out the idle mixture.

Because additional air enters the intake manifold, carburetors used with the PCV system are calibrated to compensate for the air and blow-by gas entering the intake manifold from the crankcase.

Keeping the PCV valve and the vent tube from the air cleaner to the engine clean is essential to the correct operation of the engine because this is the only crankcase venting. A connection in the carburetor base allows attachment of the tube from the PCV valve.

Because the vapors are reburned instead of escaping to the atmosphere, pollution is reduced. Because the blow-by components are positively removed from the crankcase, engine life is increased. A third benefit is that of added economy. A Chevrolet publication points out a gain in gasoline economy ranging from 2.3% at 50 MPH to 4.8% at 20 MPH and up to 15.4% at idle. This gain in economy is because the blow-by gas returned to the intake manifold is combustible and becomes fuel for engine operation.

Because current carburetors are calibrated richer to make up for the flow of air through the PCV valve, if the valve becomes plugged or inoperative even part of the time, the mixture will be much richer—with a resultant loss of economy.

The PCV valve should be checked at 6,000-mile intervals, and replaced at 24,000 miles.

EXHAUST EMISSION CONTROLS

There are various approaches to the problem of making engines and vehicles meet the emission standards. The first approaches were engine modification and air injection.

By 1972, another engine modification was being used: Exhaust gas recirculation or EGR.

Engine Modifications—Carburetors are set up with leaner or richer mixtures in the idle, off-idle and part-throttle ranges. Mechanical limiters on the idle-mixture adjustment screws prevent excessively rich idle mixtures. And, choking mixtures are eliminated very quickly after the engine has been started. Inlet air is heated to ensure good fuel vaporization and distribution.

Decelerations create very high manifold vacuums unless special controls are used. With a closed throttle, so much exhaust is sucked back into the intake manifold that the A/F mixture is diluted (leaned) to the point of borderline firing, causing missing and consequent high emission concentrations of unburned hydrocarbons. Several controls can be used singly or in combinations to control deceleration emissions:

1. Shut off the fuel flow so that there will be no unburned hydrocarbons—because there is no fuel entering the manifold. This method has the problem of creating a "bump" when the fuel is turned back on near normal idling manifold vacuum.

2. Supply a richer mixture to ensure burning.

3. Retard throttle closing to avoid high vacuum buildup. Although this latter method is commonly used, it reduces the braking effect which would have been obtained from the engine during deceleration with a closed throttle.

Distributors are set up with more retard at idle and part-throttle and various switches, valves and other controls are used to provide advance or retard as required to meet emission requirements.

Additionally, basic engine modifications are being made, including reduction of compression ratio to reduce combustion pressures and temperatures, valve-timing variations and re-design of combustion chambers. Exhaust restrictions are also being used.

CONTROLLED COMBUSTION SYSTEM

This system increases combustion efficiency through leaner carburetor adjustments and revised distributor calibration. On the majority of installations, special thermostatically controlled air cleaners in conjunction with a heated air source piped over the exhaust manifold exterior are used to maintain intake air at approximately 100°F. or above. The preheated air improves cold-start drive-aways and prevents carburetor icing during mild ambients (30°–40°F.) and high humidity. The leaner mixtures can be tolerated with warmed inlet air because the warmer mixture gives improved distribution to the various cylinders.

The air-cleaner assembly includes a temperature sensor, a vacuum motor, a control damper assembly, and connecting vacuum hoses. The vacuum motor controlled by the temperature sensor operates the damper to control air flow, providing either preheated air from a shroud around the exhaust manifold or unheated underhood air—or a combination of the two.

AIR INJECTION REACTOR AIR SYSTEM

Some engines have been equipped with AIR ever since the method was first introduced. Most of the cars equipped with these systems are those with high-performance engines. As of 1972, air-injection systems began to be used more widely to help meet increasingly stiff requirements for reduced HC and CO emissions.

This system adds air to the exhaust to continue burning of the unburned exhaust gases and reduce their HC and CO content. Air is drawn into an air pump where it is compressed and fed out through valves into the exhaust manifolds (V-8 engines) or cylinder-head exhaust ports (six-cylin-

The inside of a PCV valve has a spring and a moving part. If it rattles, it's probably O.K.

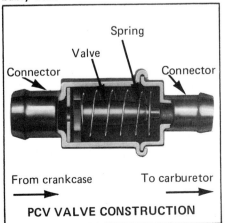

PCV VALVE CONSTRUCTION

der engines). When this compressed air mixes with the hot exhaust gases, further combustion occurs, burning most of any remaining HC and reducing CO in the exhaust before it leaves the vehicle's tail pipe. This increases the temperature of the exhaust gases.

Although this is a costly approach, the slightly richer mixture which can be used on engines which are thus equipped makes the car more driveable with less tendency toward mid-range surging.

The other side of the coin reveals a 5 HP loss to drive the AIR pump at high RPM.

Even though there is this slight engine power loss, the air pump *itself* does not significantly decrease fuel economy. For the afterburn to occur, the concentrations of hydrocarbons in the exhaust must be very high. This is accomplished by setting the carburetor significantly richer than would be required for driveability and maximum economy. This is where a lot of gas mileage has gone in recent years. The answer is not to remove, or disconnect the air pump. If this is done, the carburetor continues to dispense a rich mixture—which in turn makes the car a worse polluter than older model cars without air pumps. The only way to improve the economy of an engine equipped with an air pump is to recalibrate the carburetor so it supplies a leaner mixture.

Carburetors and distributors for engines with the AIR system are designed particularly for these engines. They should not be interchanged with or replaced by a carburetor or distributor designed for engines without the AIR system.

Exhaust-system components are also modified to get the best possible burning of any residual HC. As of 1973 vehicles, we are seeing enlarged exhaust manifolds to increase the heated area for further insurance that all possible residuals will be burned. It is expected that the exhaust-system design will play an ever-increasing role in emission reduction during the coming years.

COMBINED EMISSION CONTROL SYSTEM

In 1971 a new device was introduced on a number of vehicles: The Combined Emission Control Valve (CEC).

This emission-reduction device functions as follows.

When the valve is energized by the transmission, it acts as a throttle stop by increasing idle speed during high-gear operation of the engine. This controls hydrocarbons during deceleration by operating in the lean off-idle system of the carburetor. Keeping the lean mixture during deceleration is especially helpful in controlling HC. The limiting factor in this HC-reducing procedure is the amount of throttle opening a given vehicle can tolerate before objectionable deceleration rates occur. Do not set the CEC engine RPM above recommended specifications or your engine speed during deceleration could make braking difficult.

TRANSMISSION-CONTROLLED SPARK ADVANCE (TCS)

Transmission-controlled spark advance prevents operation of distributor vacuum advance in the lower gears. On cars with automatic transmission, TCS allows distributor vacuum advance in high and reverse gears.

EXHAUST GAS RECIRCULATION (EGR)

During the combustion process, nitrogen tends to combine with oxygen at temperatures of about 2040°F., forming oxides of nitrogen (NO_x). Engine-combustion temperatures often exceed this figure. To reduce the formation of

Engines equipped with smog-fighting air pumps mix fresh air with exhaust gases as the exhaust emerges from each cylinder. The addition of oxygen at that point promotes continued combustion in the exhaust manifold so as to burn up anything that didn't get fully burned in the combustion chamber.
The plumbing for an air-pump system looks like the drawing at left.

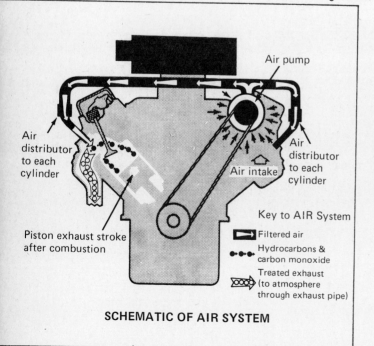

SCHEMATIC OF AIR SYSTEM

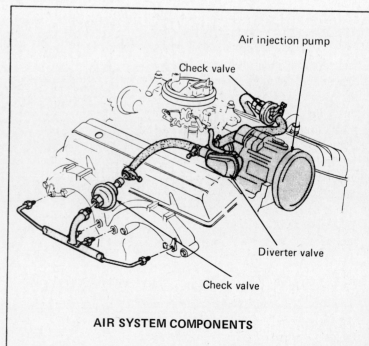

AIR SYSTEM COMPONENTS

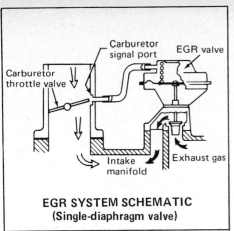

EGR SYSTEM SCHEMATIC
(Single-diaphragm valve)

Here's the basic idea of Exhaust Gas Recirculation. When the throttle is neither closed nor fully open—that is in some middle position—the EGR valve opens and allows some exhaust gas to flow into the intake manifold to mix with fresh fuel and air. It prevents formation of noxious NO_x.

A more recent development is a two-diaphragm valve which reduces engine surging at part-throttle under light loads.

NO_x, an EGR valve meters inert exhaust gases into the intake manifold to displace a portion of the air/fuel mixture in the cylinders so peak-combustion temperatures and pressures are lowered. A portion of the exhaust gas is being recirculated into the combustion chamber, hence the term *exhaust gas recirculation*.

Although EGR does not affect vehicle operation at idle or wide-open throttle (where no EGR is occurring), engines do "object" to the recirculation of unburnable exhaust gas into the combustion chambers in amounts ranging up to 15% of mixture volume. The "objection" is made known by surging and uneven running during light-throttle/high-vacuum running at speeds from 25–60 MPH.

The EGR valve mounts on the intake manifold and connects to an exhaust manifold port on six-cylinder engines or the exhaust-cross-over passage on V-8's. The valve remains closed during idle and deceleration to prevent rough running caused by excessive exhaust-gas dilution of the air/fuel mixture.

The EGR valve gets its opening signal from the carburetor. Vacuum from an off-idle port opens the valve when the throttle blade is opened past idle to uncover the EGR off-idle port.

For the engine to continue to run while the EGR system is in operation (part and intermediate throttle) the carburetor must dump a very rich mixture of gasoline down the manifold—simply to keep the fire lit—since exhaust gas doesn't burn very well. This means very poor economy or poor driveability any time the EGR system is operating. The poor driveability is manifested by spitting, coughing, or roughness during part throttle operation. Here again the problem is in carburetor metering—too rich—in order to "cover up for the EGR." Unfortunately, if the carburetor is leaned down to deliver maximum fuel economy during part throttle operation and the EGR is left intact extreme driveability problems will result—and possibly even worse gas mileage.

EVAPORATION CONTROL SYSTEM (ECS)

This system reduces fuel-vapor emissions which would otherwise vent to the atmosphere from the gasoline tank and carburetor bowl. The evaporation-control system is one of the few emission-control devices which actually adds to the efficiency of the automobile. Any time raw fuel can be burned instead of polluting the air, we are aiding car economy *and* ecology.

The obvious part of this system is the ECS charcoal canister which is usually found in the engine compartment. The ECS canister is a simple container for charcoal granules. The fuel vapor inlets and outlets are on one side of the charcoal and the purge air opening and filter are on the other. *Purge air* is the air drawn through the saturated charcoal to pick up the stored fuel vapors.

Vapors enter the canister from the carburetor bowl vent and the fuel tank liquid/vapor separator. The vapors are "stored" (adsorbed) onto the surfaces of the charcoal granules and stored in the air space in the canister. When the car is restarted, the fuel vapors and fresh air are sucked into the carburetor or intake manifold to be burned in the engine.

The ECS system saves you gas that would normally be lost out the gas tank vent. The carburetor is metered to work with the ECS system so nothing is to be gained by disconnecting any part of the

If you think trapping fuel-tank vapors in a charcoal canister is a simple proposition, take a look at this drawing. This emission device saves gas.

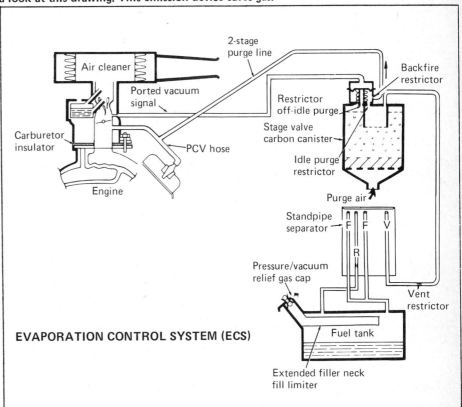

EVAPORATION CONTROL SYSTEM (ECS)

ECS. Severe cold start driveability problems can be incurred if the system is not allowed to use the fuel in the canister during start-up.

The air filter at the bottom of the canister requires replacement at 24-month or 24,000-mile intervals.

Certain wheel-splash conditions can plug the air filter in the vapor-canister base prematurely. The filter must pass air for the system to function. Inspect it periodically for mud and other contaminants.

ECS also uses sophisticated measures in the fuel-tank filling and construction areas. Thermal expansion is provided for by trapping up to three gallons of air during tank filling, then allowing this air to escape to the top of the main tank where it vents to the canister through a liquid/vapor separator. Any liquid trapped in the separator/s drains back to the tank and the vapor passes to the canister.

The gas cap is designed to contain the vapors in the tank and thus force them through the separator and into the canister. It takes approximately 1/2 PSI to make the canister system work. Under extreme conditions, when the tank pressure goes above cap limit it starts venting to atmosphere. First-system caps would blow off at 1–2 PSI pressure; varying by make and model. Starting in 1972 the fuel-tank caps have a higher blow-off pressure setting so the tank, straps and related components are stronger to withstand the increased pressure. The caps also include a vacuum vent to release any vacuum which forms in the tank as the fuel is used.

Do not mix caps between various makes and models. With large surface areas existing in fuel tanks it doesn't take much pressure to damage or burst them. Any emission parts which fail should be replaced only with original replacement parts to ensure correct performance.

CATALYTIC CONVERTERS

Faced with emission-control laws that require less emissions from future car models, auto manufacturers reached a point of diminishing returns with add-on devices and fixes.

Re-thinking the entire problem led to use of catalytic converters in the exhaust system. These converters require no-lead fuel and a few tanks of leaded fuel will ruin them. When properly used—with no lead gasoline—the converter takes out much of the pollutants in the exhaust.

The payoff is that engines can be built with more nearly normal carburetion and ignition. Many of the add-on fixes can be left off. The engine is more like the good-ole days and has zip along with better fuel economy than other smog-controlled engines.

Disadvantages include high heat under the floorboards, higher cost of no-lead gasoline, and the fact that these converters reduce conventional emission problems but cause a new one—sulfuric acid comes out of the tailpipe.

Many Detroit cars used catalytic converters in the 1975 model year, amid great controversy about whether they are an improvement or not. The answer has not been determined at this time.

EMISSION SETTINGS

Currently all manufacturers are required to install a specification label in the engine compartment. It details the correct carburetor and distributor adjustments needed to maintain legal emission levels for that vehicle. No matter how effective or sophisticated the control system, a weak spark, fouled plug, bad plug wire, cracked distributor cap—or any of 200 other things—can wreck the system's efficiency. It only takes one of these things going wrong to cause the car to become a pollutant emitter of a worse nature than a car with no controls at all.

A single bad spark plug is a good example. In SAE Paper 710069, "Exhaust-Emission Control for Used Cars," the authors pointed out that a single fouled plug increased the HC emission level six times.

WHAT'S IT COSTING?

The whole idea of emission controls is to clean up the environment—namely the air we breathe . . . and live in. But, as is always the case when you get down to the facts—there is "no free lunch." Anything you get—in this case cleaner air—costs *something*. That something includes reduced driveability, increased gasoline consumption, increased tune-up costs—and a car that weighs more.

To these obvious costs we must add increased complexity of the entire vehicle with associated cost increases in both engineering and manufacturing.

In 1975 many manufacturers began use of catalytic converters in the exhaust system to meet tighter emission regulations. These get very hot and require special safety equipment to shut them down if they overheat to the point of being dangerous. During the 1975 model year, it was found that these converters produce sulfuric acid which goes out the tailpipe. It's not clear if the authorities prefer conventional pollutants or sulfuric acid in the air we breathe, so the future of the catalytic converter is in doubt.

PERFECTUNE ROCHESTER CARBURETOR KITS

Explaining what the Perfectune Kit *is not* may be easier than explaining what it is. It is not a fly-by-night gimmick which promises to give great increases in fuel economy or great increases in horsepower. The Perfectune Kit is an engineered package of metering rods and jets with detailed instructions on how to tune a Rochester Q-jet carb for better mileage, driveability or performance.

If you are after better mileage, you can expect 1 to 5 MPG increase. If you are after a carb rebuild kit, you should buy a carb rebuild kit—because the Perfectune kit is not that. It contains parts not found in a carb rebuild kit. Simply stated, the Perfectune kit allows a Q-jet to be tuned for a specific altitude, work load, driving habits and several other factors. It allows precision tuning for your driving situation which can not be done at a factory building hundreds of thousands of cars.

What are the drawbacks? The most obvious one is that the Perfectune kit is only for Q-jets—which are found on most full-size GM cars. If you own a Ford or Chrysler product, the Perfectune won't do you any good because it won't fit. The second drawback of the Perfectune kit is that the carburetor must be partially disassembled for the installation. This is no big deal with the excellent instruction booklet provided. No special tools are required—but the carburetor must be opened up.

An increase in fuel economy of 1 to 5 MPG is certainly worth the effort—although not spectacular. The simple fact that the kit delivers what the advertising claims is spectacular enough. Write to Perfectune Kit, 4230 S. 36th Place, Phoenix, AZ 85040.

SMOKE OUT THE EXHAUST— WHAT DOES IT MEAN?

When you're following a bunch of vehicles and one of them is in the process of starting to pass and a cloud of smoke rolls out the exhaust, you know two things. First, the driver has buried his foot in the throttle. Second, smoke out of the exhaust means poor economy.

If the car is an older one the problem could be oil consumption, but speaking primarily of a new vehicle in obviously new condition, it is likely the air/fuel calibration is not correct.

A car that belches great amounts of smoke in a passing maneuver may have incorrect carburetor calibration. Probably the majority of the vehicles driven today are in that category. The situation becomes far more extreme as you increase altitude. Any time the power system in the carb is pulled in, the calibration is so far out that the car will never run well. Down at sea level even with these great clouds of smoke emitting from the exhaust, the car did indeed move out and respond—perhaps not the best it could but certainly close to its full potential. At altitude there are many factory-calibrated cars that will not even run well. This mismatch of fuel/air ratios in the carburetor will hardly pull the car. For sure the cars won't pull a trailer and they'll hardly maintain speeds in top gear at higher altitudes. In short, they are sick. This is a real shame and doubly so because people become unwitting victims of this poor performance. But in the competitive world as it is, one doesn't have to do any better than the other so the calibration of all cars is approximately the same . . . . very, very poor at wide open throttle in altitudes.

What can you do about it if the problem belongs to you? It depends. If you own a late model GM car equipped with a four barrel Quadrajet carb, read about the Perfectune kit. If you own a car equipped with some other make of carburetor, write to Perfectune, state the make and year of your vehicle. Perhaps they will have a solution for your car.

If you have a GM car equipped with a one-barrel (Monojet) or two-barrel (2G, ZGC), H. P. Books' *Rochester Carburetors* book has complete information on modifying these carbs for better economy.

137

9 LEARN TO TUNE-UP YOUR CAR

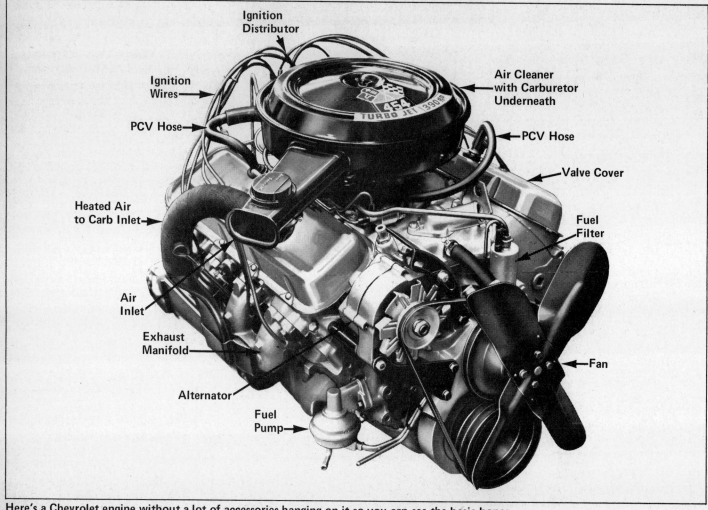

Here's a Chevrolet engine without a lot of accessories hanging on it so you can see the basic bones.

Probably there are more people tuning their own automobiles than you think. A major part of the automotive aftermarket is sales of parts and tools to the home do-it-yourself tuner.

People who spend an occasional Saturday morning servicing and tuning the car usually don't talk about it much because they think it's an ordinary thing to do and everybody does it.

People who never lift the hood also think that's the normal way to live—so they don't talk about it either. However, a steady flow of converts to the greasy fingers group is taking place partly because of the high cost of having someone else do it for you.

Raising the hood of a modern automobile with the intention of doing something serious under there requires a good supply of self-confidence and mechanical bravery—the first time you do it.

It helps a lot to see the procedures and observe a lot of the little details that are difficult to put on paper or capture in a photo. That's why I commented earlier that lots of people tune their own cars and don't talk about it. You probably know people who do it—people where you work or somebody in your neighborhood.

Try casually mentioning that you intend to start tuning your car but have never done it before. There's a good chance you can get somebody to volunteer to help you the first time—or at least let you come over to help him work on his car. Either way, you get past the hurdle of doing it the first time. You see, it isn't really so difficult and the improvement is worth while.

TUNE-UP PROCEDURE

Tune-ups are a rather controversial subject—no one has the same idea about what needs to be done. What one man will call "minor," another will term "major." Simply stated, a tune-up is routine maintenance and the process of restoring factory specifications to an engine. How much you should do and how often will always be debated.

I'll touch on the basics here and suggest, if you are really serious about doing your own tune-ups, you acquire a factory service manual for your vehicle and refer to it for specific information. A tune-up is not a cure-all, but can provide you with many miles of trouble free, economical transportation.

Do you feel there is a specific problem with the engine, or will the tune-up be routine? While some problems will be corrected as a matter or course, others may not and you'll have to go after those separately. For instance, changing the fuel filter inside a carburetor probably wouldn't be part of your normal tune-up. But if the car has been coughing and running erratically for several days you might suspect a plugged filter and thus it could be the real reason behind performing a tune-up.

Every tune-up should begin with the battery. The battery is like a heart—attend to it first. Check the water in the battery as the first step. Replenish with distilled water only.

Dirt and corrosion on the top of the battery and on the terminals can cause current leakage. Remove the corrosion with a solution of baking soda and water—about one heaping tablespoon of soda per quart of water. This neutralizes the build-up caused by acid. Be careful not to get any of the baking soda solution down in the cells. After the terminals and battery top are clean, rinse with clear water. Dry the battery off with throwaway paper towels and check the hold-downs to be sure they hold the battery securely in place.

Remove the battery cables from the battery and clean both cable clamps and the battery posts with a wire brush or sandpaper. After replacing and tightening the clamps, coat the terminals with light grease to prevent the next build-up of corrosion.

If you have doubts about the condition of the battery, have it checked with a temperature-compensated hydrometer at a local service station. If water was just added to the battery, you should allow at least one day of normal driving before making this check. The specific gravity of each cell should be 1.275 to 1.280 or 1.260 for Sta-full types—corrected to 80 degrees F. Readings below this can indicate possible charging system trouble or a partial electrical short in the wiring. A gravity variation of .025 or more between cells indicates possible cell trouble.

If the battery was partially discharged as indicated by a low gravity reading, it should be recharged before being returned to service. The cause should be found and corrected. This is very important in cold climates because a battery is only about 40% efficient when the temperature gets down to zero.

PLUGS FOULED— MILEAGE GONE? YOU BET!

Everyone knows that for top gas mileage the ignition system of a car should be in top working order. Never mind the gimmicks and gadgets—how much does one or two fouled plugs affect gas mileage? We wondered too, so we rigged up one of our test cars—our '69 Pontiac V-8 so we could measure fuel economy with one plug fouled and then with two plugs fouled. The results are rather startling.

With one spark plug not firing, the test car lost a minimum of 12% of the baseline fuel economy at steady speed running. That's a lot of fuel!

With two plugs not working, gas mileage was off about 30%! That's slightly more than five miles per gallon at a steady 60 MPH. Driving a car with one or two plugs out of order (or plug wires) is making the rest of us pay part of your bill—in increased emissions. Independent testing reported to the Society of Automotive Engineers shows that on an older car without emission controls one fouled plug will increase hydrocarbon emissions by six times. On "clean" emissions engines, one or two fouled plugs would have a much more disastrous effect on hydrocarbon emissions.

The performance loss and mileage sag caused by inoperative plugs is even more dramatic in our performance running than in steady state. It took 47.8% more gas to go from 30 MPH to 80 MPH with two plugs fouled than it did making the same run with all eight firing! Never mind how much longer it took for you to make the run—which could be downright dangerous in a passing maneuver—look at all the gas you're wasting.

The message is clear—bad plugs mean poor mileage. Keep your car in tune!

What this is gonna do approximately day after tomorrow is not start your car. It's pretty certain the fluid level in this battery has not been checked for months. When fluid is low, battery power is also low.

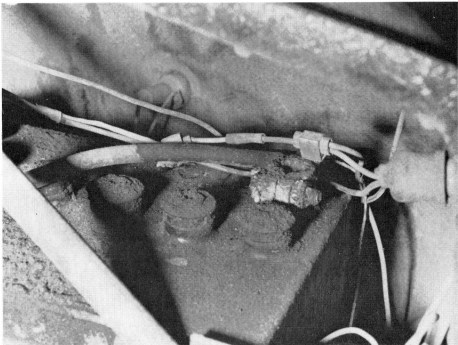

139

1/Start your tune-up by servicing the battery. If there is corrosion, mix some baking soda with warm water.

2/Remove corroded hardware and dump it in the soda and water mixture.

3/Let it bubble and fizz until it stops.

4/Rinse off the parts with clean water and let them dry in the sunshine.

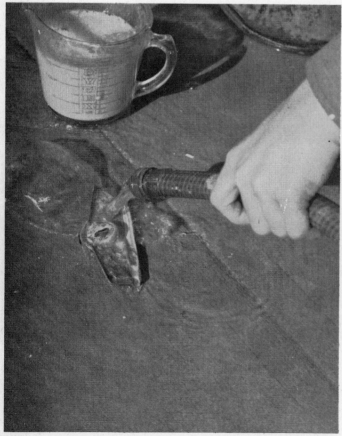

5/When dry, give them a light coating of grease or petroleum jelly to protect them from future corrosion.

6/If there is evidence of corrosion on the battery terminals, remove the cables and dunk the ends in your soda-water mixture. Clean inside of wire terminals and outside of battery posts with steel wool or sandpaper and re-assemble. Coat terminals all over with grease. If cable insulation looks swollen near terminal, corrosion has probably worked its way up inside the cable. You may as well replace it, and keep the terminals greased next time.

At this point in a tune-up, I pull all of the plugs. Put them on a work bench of someplace out of the way. The important thing is to keep them in the order they came out of the engine. If you have a V-8, make two rows of plugs. Normally, I don't regap the plugs or install new ones at this time. Just pull the plugs and leave them out—but in order. By pulling the plugs and leaving them out, you are making less work for the battery when you install points and set ignition-point dwell.

At auto parts counters you can sometimes buy just a set of ignition points for the make and model of your car. More often you will have to buy an ignition tune-up kit in a box which contains points, a new condenser, and sometimes even a new distributor rotor. The last two parts may not be needed at every tune-up but if you have to pay for them you may as well use them.

The condenser is a small cylindrical can and you just substitute it for the one you find in or on the distributor. The distributor rotor just pulls off the shaft.

When installing points you can leave the distributor in the engine or take it

Many home mechanics have tried pliers and pipe wrenches on spark plugs and then discovered this. They declare it's practically perfect for the job. It's called a spark plug socket. A ratchet handle and possibly an extension complete your standard expert's kit.

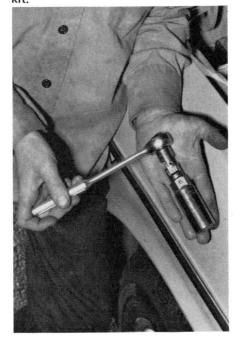

out for servicing. In some engines, the points can be checked and replaced by leaving the distributor in place. In other engines, it is practically impossible to inspect the points much less replace them. If the distributor is to be removed, *be sure to mark* the relative positions of the rotor and housing with respect to the engine so the distributor can be returned to the proper position. See the shop manual for procedure.

Points are generally replaced about every 10,000 miles. Under abnormal conditions, they may have to be replaced much sooner but it is not unusual to get far more than 10,000 miles from a set of points. The need for replacement is indicated by an electrical test or by visual inspection. The normal appearance for points is an even grey surface. Dark blue or even black areas on the points are an indication of burning. They are on their way to failure.

Install points before replacing the plugs. This way, dwell can be set by cranking the engine with little strain on the battery.

Most points sold now are packaged as assemblies and are pre-aligned—this means the contacting surfaces are parallel. New points should always be checked to see that they are parallel before being installed. If not, slight bending of the arms will do it.

The breaker cam should be checked for scoring or roughness by feeling it with your finger. This condition will cause rapid wear on the rubbing block of the points—in turn changing dwell angle and timing. Lube the cam and the rubbing block very lightly with a high quality white grease (not wheel bearing grease). The preferred way of setting the breaker point is with a dwell meter, but they can also be set with a feeler gage.

The idea of setting point gap or dwell is simple. The ignition points are opened and closed by a rotating cam in the distributor. Therefore they spend some of the time open and some closed. When the points are closed, current is flowing in the ignition coil and the coil is getting set to make a nice big spark when the points are next opened. However, it is important to leave the points closed *long enough* for the coil to get set to do its trick. It has to build up a magnetic field around the coil windings before it can make a big zapper of a spark.

The "old fashioned" way is to measure

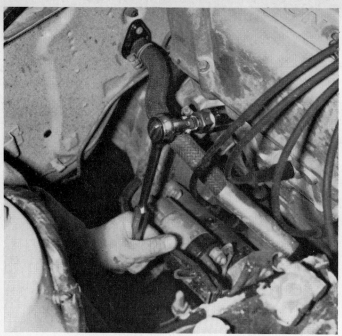

Remove plug wires by grasping the metal fitting on the end of the wire—not the wire itself. You can grab the metal fitting right through the rubber boot on the end of the plug wire. Put the socket over the plug and turn it counterclockwise. When re-installing plugs, get them snug but never as tight as they will go. Stop tightening when it is plain you could still tighten some more.

When removing plugs, keep them in order so you will know which plug came out of which cylinder. Plugs that have seen a lot of service really open up the gap. Notice particularly the gap on plug #1 at the far right. Plug #3 may show signs of oiling. You should run a compression check to see if it tells you of a problem. No, the thumb doesn't hurt much now. Pretty tender for a while though!

Before you start a tune-up buy an ignition repair kit with a new set of points. If you don't use it this time, save it for next time.

Wires just pull up and out of holes in top of distributor cap. Before removing any, tag them or memorize the pattern so there is no possibility of putting them back wrong. Flip back holding spring clips and lift off cap. Inspect inside for dirt, cracks, or worn terminals such as shown here. Rotor is not supposed to touch terminals. If there is some evidence of touching and wear, new parts may fix it or you may need qualified help to install a new distributor.

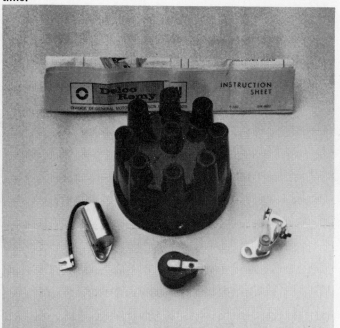

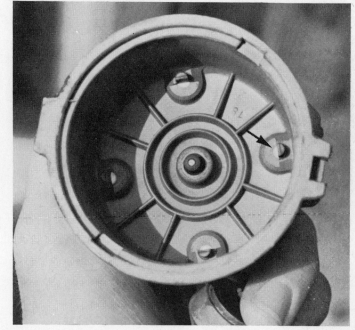

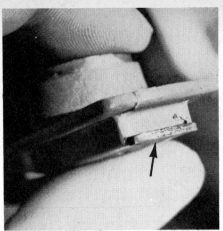

If rotor tip shows scratches, dirt or wear, same deal. Rotor pulls up and off shaft. Normally there is a groove so you can't get it back wrong, provided you shove it down on the shaft fully.

You may be able to set point gap without removing the distributor. If you are installing new points, it's handy to remove the distributor. If you do remove it, you must mark or carefully observe and remember the position of the distributor in the engine. The rotor will turn while you have the distributor out and you must put it back the way it came out! The bottom of the distributor often drives the oil pump. Be sure the distributor goes all the way back into the engine block exactly like it was before you pulled it out. Distributor is held in place by screw through oblong slot on that metal tang near bottom (arrow A). Remove screw, pull distributor straight out. After installing new points, put a light smear of high-temperature grease on the cam lobe of the rotor (arrow B).

Buy a set of feeler gages if you intend to set ignition points by measuring the gap. This set works for plugs and points.

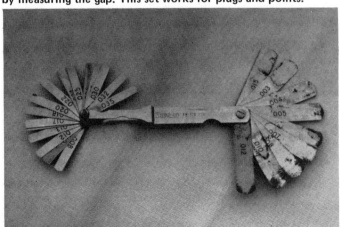

Then use them to measure the gap.

the gap between the points when they are opened to the maximum amount by the cam. Basically, the amount of gap tells you how much of the cam profile was used in holding the points open. If the cam opens the points and holds them open for a long time, it must also move them open a relatively long distance. On the other hand, if the cam just barely gets the points open before they close again, not much of the cam profile or rotation was used up in holding the points open.

The important thing is not how much of the time the points are open, but how much they are closed. But whenever the points are not open, they must be closed, so it doesn't matter which you check.

If you measure point gap when fully opened, and it is within spec, then the length of time the points are closed should be correct.

To measure gap, you need a set of feeler gages which you can buy at any auto parts store or the tool department of a large department store. These are flat strips of metal, all mounted on a post at one end, so they fan out like a deck of cards. Each is marked with its thickness and the range of thicknesses will run from around 0.001" to around 0.40".

To find out how much gap there is, put feeler gages into the gap between the points until you find the one that just fits snugly with a little friction on both sides of the gage as you insert it and draw it back out.

Rotate the distributor cam until the rubbing block sits right on the tip of the cam to be sure you are measuring at *maximum* gap. Check it again.

Rotating the distributor cam is easy with the distributor out of the engine. If you prefer to leave it in, you can rotate the engine by a tug on the fan belt if you have already removed the spark plugs. If you are skillful, or lucky, you can get the distributor cam to stop when the points are fully open.

If the gap is not correct you have to adjust it. The point set or point assembly is a complete unit with both points and a pivot post all on a common base plate. The base plate is attached to the distributor by a screw. Loosen that screw and rotate the entire point set so it sits closer to the distributor cam or farther away from it. Moving it closer to the cam will cause the point gap to be wider.

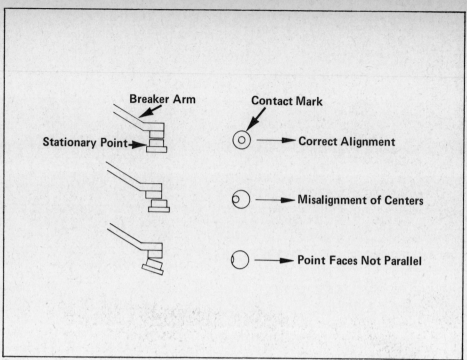

Things to check on points are the alignment between the two contact surfaces by examining them visually and the shiny "contact mark" on the surface of the points which shows where they have been touching each other. When the points are centered and parallel, the contact mark is centered as shown. Misalignment of centers or faces not parallel give different indications.

Sometimes a screw adjustment changes point gap. Sometimes a screw with an eccentric head will move the point set as you rotate the screw. The manual shows you how.

Set point gap to spec and tighten down the clamping screw. Measure the gap again.

Point dwell is an electrical measurement of how much time the points stay closed—how much time they *dwell* together. It's done with a special dwell meter. Hook up the meter according to the instructions that come with the meter or the shop manual and the meter pointer will show you a dwell number in degrees. If it isn't right, you change it by altering the point gap just as you do to set point gap. Dwell and gap basically mean the same thing to the engine.

Dwell is usually specified as a range—for example 28 to 30 degrees.

To measure dwell with a dwell meter simply clamp the wires from the instrument where the instructions tell you and start cranking the engine—with the plugs out. Normally, one of the wire terminals will clamp to the distributor side of the coil and the other is grounded to some part of the engine. What must be kept in mind is that changing dwell changes ignition timing. Increasing dwell retards timing; decreasing dwell advances timing. Therefore you should *always* set dwell or point gap before setting timing. No matter which method you use to "set the points" always recheck your adjustment (either gap or dwell) after cinching down the lock screw. Reset if it has changed.

Most distributors have two advance mechanisms—a centrifugal mechanical advance for controlling ignition timing at various speeds and a vacuum advance for more advance for when the engine is not working hard. These mechanisms can be checked with the distributor in or out of the car by using special test equipment. A quick check without special equipment can reveal some problems. To check for a sticking mechanical advance with the distributor removed, hold the shaft to keep it from turning, grasp the rotor and turn it in the direction of normal rotation. You should feel some spring tension and when

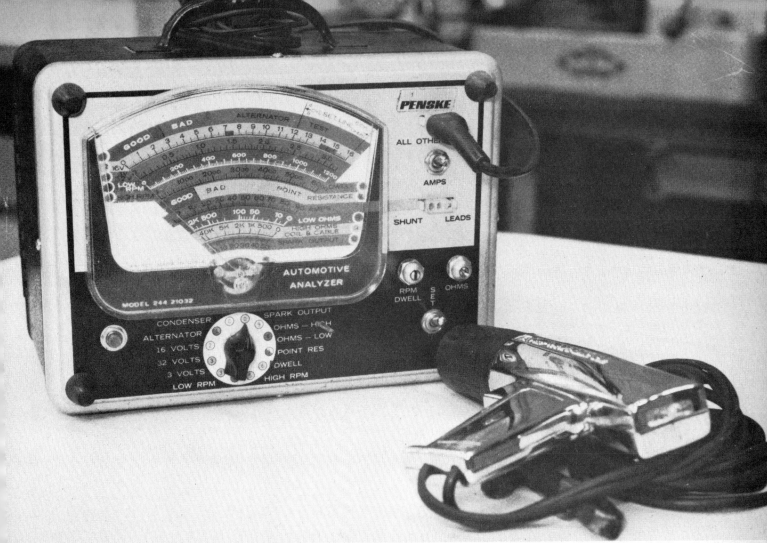

Test equipment like this gives you a professional tune-up and pays for itself while doing it. Test meters measure a lot of different things and go under a lot of different names such as "Engine Analyzer" and "Automotive Analyzer." One selector switch position will set it for measuring dwell. Hook it up according to instructions. The other "tool" is a timing light.

Point dwell is easily adjusted on GM distributors with a hex-wrench, through a little trap door as shown. The shop manual gives more information on this exact procedure.

145

you let go of the rotor, it should return to its original position. If it doesn't return, the advance mechanism is sticking. With the distributor installed in the engine, you can just twist the rotor and observe the advance and return.

Vacuum advance chambers sometimes leak—and while this might not affect performance, it could be responsible for poor gas mileage. To check for a leaking vacuum advance chamber, disconnect the vacuum line to the chamber before the distributor is removed from the engine. Move the breaker plate or the distributor housing to the maximum advance position to compress the spring in the chamber. Press the tip of a finger over the inlet to the chamber and release the breaker plate. The vacuum should hold the breaker plate in about mid-position. If there is a leak, the breaker plate will slowly return to its original position.

Check the inside of the distributor cap for any oil or dirt. Clean with a dry cloth. Inspect for very fine, hairline cracks or carbon tracks between the electrodes. The presence of such cracks renders the cap unusable. Remove each plug wire from the cap (one at a time) and inspect for signs of corrosion. If not corrected, this corrosion can cause an ignition miss.

The only wear in the rotor is at the tip. If this appears worn and eroded the entire rotor should be replaced. Don't attempt to file off the rotor tip because this creates an extra wide gap.

Inspect the spark plug wires for brittleness or cracking or other signs of deterioration. Most vehicles use a resistor type of wiring instead of a solid wire in order to minimize radio and TV interference. These wires require more frequent inspection.

If you had the distributor out of the engine, you can put it back now. This is when you are glad you marked or noticed the position of the distributor rotor in relation to the body of the distributor. Leave the cap off so you can watch the rotor as you reinstall. Also, you should have marked the distributor body in reference to the engine block so you can drop it back down into the hole in exactly the same rotational position it had when you took it out.

There will be a tang or a small gear on the bottom of the distributor shaft which must mate with a part of the inside of the engine. When the parts are mated, all

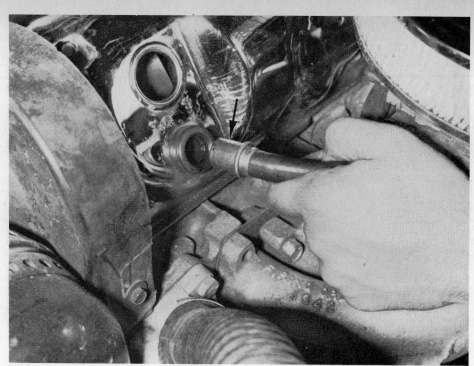

While you're in the neighborhood, locate the PCV valve which may be in the end of the hose.

Or else the PCV valve may be in the middle of the hose.

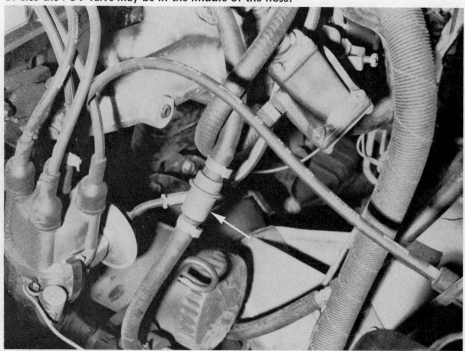

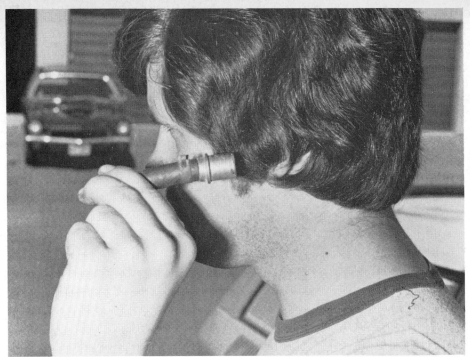

Wherever you find the PCV valve, remove it and shake it. It should rattle. If not, soak it in solvent until it does. If not, replace with a new one.

To check cylinder compression, you need a gage. This one is threaded to screw into a spark plug hole, which is mighty handy. The instructions that come with the gage, or a shop manual, will describe other tests besides the one covered here.

your reference marks should line up again—otherwise you have lost the basic timing of the engine.

Tighten the clamp bolt to hold the distributor in place.

Occasionally, when you are servicing the engine, you should make a compression test using a compression gage. This is an indicator that fits into the spark plug hole. When you crank the engine, the gage records the amount of pressure the engine was able to produce in that cylinder. Without the proper amount of compression an engine can never perform fully and you're wasting your time trying to cure the problem with plugs, points and setting timing. If the compression is low it means something is wrong inside the engine. Maybe something is broken—or maybe everything is just wearing out.

Compression testing is simple but pretty easy to do incorrectly. It is important to understand you are comparing the compression of each cylinder with all other cylinders in the engine. That means you must test every cylinder in exactly the same manner. If you are doing the testing alone, you'll have to connect a remote starting button to the starter solenoid so you can crank the engine while holding and reading the compression gage.

Some gages are threaded to screw into the plug hole. If you have the type with a rubber cone on the end, push it down hard into the spark plug hole—then give it half a turn. Continue to shove hard on the gage while cranking the engine for five full compression strokes. Discerning the compression strokes will be the easy part. Note how the pressure builds on each stroke. At the end of the last of five compression strokes note the final pressure of that cylinder. Write it down and then move to the next cylinder. Make sure you record the reading of each cylinder after it is tested. A service manual will tell you what the compression should be—or the manual will indicate the allowable difference between high and low cylinders. The maximum variation will be given as so many pounds per square inch—such as 10 PSI. If all the compression readings are within tolerance, you can assume the inside of the engine is in reasonable condition. If one or two cylinders give a much lower reading than the other, I suggest rechecking the cylinders. If the readings still come out low, take the car to a mechanic.

With or without the compression check, you are now ready to install new or the old spark plugs. All of the old plugs should be in roughly the same condition. If the insulator tip is a normal light brownish color on some plugs, but oily and carboned on others, spend a little time looking at the plug wires, terminal ends and distributor cap. It is pretty evident you have some plugs that are not firing every time they should.

Spark plugs do wear out but they can

With the plugs removed, you have to rotate the engine using the starter motor to make a compression check. If you have an assistant, use the starter switch in the driver's compartment. If you work alone, you need a remote starter switch like this.

normally be cleaned and regapped several times before having to be replaced. I soak mine in benzine overnight. A druggist can order the cleaner for you if he doesn't have it in stock, then carefully scrape away as much of the residue around the tip of the plug as I possibly can with a very small file or little piece of wire.

Use a spark plug gapping tool—$1 or so at most auto parts stores—restore the correct gap of the plug by bending the side wire as needed. Reinstall the plugs. If you are installing new plugs, make sure you check the gap before installing them. Coming gapped from the factory does not insure the plug has not been dropped or hit hard enough to alter the gap.

There are two types of spark plug seats—the kind that uses a washer-gasket—and the kind that does not. The kind that does not is called *taper seat*. You just screw these in the spark plug hole.

Unless you can plainly see that the base of the plug is tapered where it fits against the metal of the engine, the plug uses a gasket. Usually the gasket will be installed on the plug and crimped so it can't fall off. Sometimes the washer will be loose in the package with the plug. Find it and use it.

Tightening spark plugs properly should be done with a torque wrench which measures the amount of tightening force you are using. However, about 99% of all mechanics and 101% of home mechanics do not use a torque wrench. You can damage the plug or the threads in the engine by using too much force when installing a plug.

Most old hands have learned the "feel" of it by experience and you can learn it by starting with a new plug and gasket or at least a new gasket. Turn the plug in until it is snug, or finger tight, but you haven't tightened it yet. Then turn the plug about 1/2 or 3/4 of a turn more and stop. It will be plain that you can turn it some more, but don't. You want just enough force to compress the gasket. If you notice it, you can actually feel the gasket compressing during that last 1/2 turn of tightening.

Many plugs today do not have a gasket. When this type is being tightened, screw them in until the plug is snug against the seat of the plug hole. Now tighten it lightly but positively about 15 to 18 ft./lb. torque.

Screw in the compression gage—or hold it in place if it's the other kind—crank the engine and note the gage reading. Do it on each cylinder.

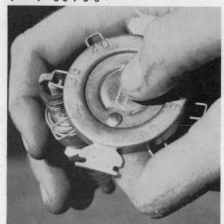

Whether plug is old or new, check gap against specification by using round-wire spark-plug gap gage.

If gap is not correct, make it so by bending side electrode. Spark plug gages have a special tool attached for that purpose.

How to read spark plugs
Courtesy of Champion Spark Plug Company

Scavenger Deposits—Fuel scavenger deposits may be white or yellow in color. They may appear to be bad, but this is a normal appearance with certain branded fuels. Such materials are designed to change the chemical nature of deposits to lessen misfire tendencies. Notice that accumulation on the ground electrode and shell areas may be unusually heavy, but the material is easily flaked off. Such plugs can be considered normal in condition and can be cleaned with standard procedures.

Worn-Out—Worn-out, eroded electrodes and a pitted insulator are indications of 10,000 miles or more of service. For better gas mileage, quicker starting and smoother engine performance, replace plugs when you see these conditions.

Oil Fouled—Wet oily deposits may be caused by oil leaking past worn piston rings. "Break-in" of a new or overhauled engine before rings are fully seated may also produce this condition. A porous vacuum booster pump diaphragm or excessive valve stem guide clearances can also cause oil fouling. Usually these plugs can be degreased, cleaned and re-installed. While hotter type spark plugs will reduce oil-fouling, an engine overhaul may be necessary to correct this condition.

Carbon Fouled—Carbon fouled plugs show dry fluffy black deposits which may result from over-rich carburetion, over-choking, a sticking manifold heat valve or clogged air cleaner. Faulty breaker points, weak coil or condenser, worn ignition cables can reduce voltage and cause misfiring. Excessive idling, slow speeds under light load also can keep plug temperatures so low that normal combustion deposits are not burned off. In such a case a hotter type spark plug will better resist carbon deposits.

Splash Fouled—This can occur with relatively new plugs and often immediately after tune-up. Accumulated combustion chamber deposits are melted and thrown against the plugs at high engine RPM. These plugs can be cleaned and re-installed. Malfunction should not be attributed to the plugs since they are victims of the operational environment.

Burned Electrodes—Burned or blistered insulator nose and badly eroded electrodes are indications of spark plug overheating. Improper spark timing or low octane fuel can cause detonation and overheating. Lean air fuel mixtures, cooling system stoppages or sticking valves may also result in this condition. Sustained high-speed, heavy-load service can produce high temperatures which require use of colder spark plugs.

Preignition—If one or two plugs of a set have melted electrodes, preignition was likely encountered in those cylinders: Check for intake manifold air leaks, possible crossfire, or worn distributor parts (lobes-bushing) which might alter cylinder to cylinder timing. Be sure that plug is of proper heat range. Find the cause of preignition before placing the engine back in service or further damage could result. Engine may already be damaged inside (melted piston crown) when you find a plug looking like this.

Reversed Electrode Bend—The improper use of pliers-type gap setting tools will bend the side electrode and frequently push the center electrode into the insulator assembly. Because of the force multiplication exerted by these tools, they must be used with care.

Broken Insulator—If one or two plugs of a set have fractured insulators, severe detonation is suggested. Indiscriminate bending of the center electrode during gapping can also cause the nose cracking. Check for cause of detonation and replace with new Champions correctly gapped and of proper heat range.

Reversed Polarity—Polarity refers to the direction of spark travel. It should always be negative, meaning that current should arc from center to ground electrode. Reversing the flow (ground to center) places greater demands on ignition output by increasing the required firing voltage. Symptoms of reversed polarity are: 1/hard starting, 2/missing during acceleration or at higher speeds, 3/short spark plug life.

Mechanical Damage—Bent electrodes and a broken insulator result from some foreign object falling into the combustion chamber. Because of valve overlap, such objects can travel from one cylinder to another and damage can occur accordingly. Cylinders should always be cleaned out to prevent recurrence. Such damage can also be caused by installing a plug of longer reach than required by the cylinder head resulting in mechanical contact between the firing end and moving engine parts. Care should be exercised during carburetor replacement or overhaul to prevent any metallic objects from inadvertently entering the intake manifold.

Normal—Normal plugs have brown to greyish tan deposits and slight electrode wear, indicating correct spark plug heat range and mixed periods of high and low speed driving. Spark plugs having this appearance may be cleaned, re-gapped and re-installed. This was the good news. Most of the rest of these photos show bad news of one sort or another but with proper care, it won't happen to you. All spark plug photos in this section are courtesy of the Champion Spark Plug Company.

At this point, with the spark plugs re-installed, the engine should run but maybe not well because when you set point gap you also changed ignition timing.

Anytime the distributor is removed or new points are installed (regardless of whether the distributor has been removed from the engine) the ignition timing should be checked. The same holds true whenever the dwell angle is changed, whether the points are old or new. Timing is accomplished either with the engine stopped or running, depending on the recommendation of the manufacturer. The most common procedure is to time the engine while it is running (normally at idle). It is important the engine be operated at the speed specified in the manufacturers tune-up specifications. It is usually recommended that the distributor vacuum line be disconnected and plugged during timing.

Next step is to set timing with a timing light while the engine is running. The timing indicator is normally a pointer on the front of the engine block and a groove or some lines on the vibration damper which is driven by the crankshaft and is normally the lowest "pulley" on the front of the engine.

It helps to wipe the vibration damper clean with a rag before starting the engine because if it's dirty the timing marks are hard to see. Some mechanics fill the lines by rubbing across them with a piece of white chalk to make them more visible in the timing light flashes.

The idea is that spark plug #1 is supposed to fire just when the mark on the damper is opposite the pointer on the engine, or at some other definite relationship between damper marks and pointer. More on that in a minute.

Having located a pointer on the block and marks on the damper, the next problem is to identify spark plug #1 which fires cylinder #1. Don't laugh. Don't guess. Sometimes the cylinder numbers are marked on the engine. If not, check the manual. If you try to time off the wrong spark plug you won't succeed.

A timing light is a special flashing light usually shaped like a pistol with some wires coming out of the handle. One of the wires connects to spark plug #1 with an adaptor so you can connect both the timing light wire and the normal ignition wire to that plug.

Other wires from the timing light con-

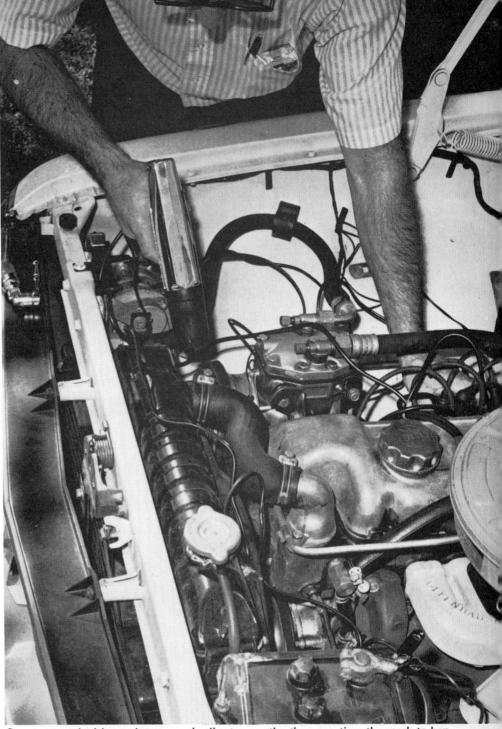

Once you get ignition point gap or dwell set correctly, then you time the spark to happen at the proper instant. Hook up your timing light according to instructions. The light flashes each time spark plug #1 fires and makes all the rotating parts appear to stand still. Be careful. They are turning and the fan can cut off your finger or wind you up by the necktie anyway.

nect to the battery and the engine. Follow the instructions for the timing light.

When you get it hooked up correctly, the timing light will make a bright flash every time plug #1 fires. For that to happen, the engine must be running. Fire it up. Be sure it's in neutral. If you want to, put some blocks in front of the wheels so the big brute can't smash you against the garage wall if it decides to go driving while you are tuning. Only the uninformed will smile.

The timing light works like a strobe light. It flashes when spark plug #1 fires. Point it at the vibration damper and pointer and it will illuminate them each time plug #1 fires, you will see it happen.

CAUTION: The engine is running but the strobe action of the timing light may cause it to look like it's not. The fan is turning and the fan belts are all anxious to grab you by the sleeve or necktie and cause you pain and agony. Keep fingers and clothing away from the moving parts.

Usually there is a scale of some kind on the damper with some lines. On one side of the zero mark the scale will be marked with an A or maybe a +. On the other end of the scale it will be marked with an R or maybe a -. A or + means the spark is advanced too much, happening too early, has too many degrees of advance. Check the manual. It should show you how to read the pointer against the scale and what the reading should be.

If the reading is not right, loosen the clamp bolt on the distributor housing and turn it slowly while watching the scale on the damper. Set it right. Tighten the clamp bolt on the distributor. Switch off. Disconnect.

You have just got yourself an improvement in gas mileage and vehicle performance of "up to" 130%.

The test lead from the timing light hooks onto the #1 spark plug wire. Most timing lights come with an adaptor to allow making the connection and also delivering spark to the plug. If you lose it, you can make one like this.

With the engine running, the timing light flashing, and your necktie not being wound up by anything, this is what you look at. There will usually be a pointer on the front of the engine with some kind of markings. On the crankshaft pulley or vibration damper will be another mark or marks. These two sets or marks should come into some particular alignment as described in a shop manual for your car. If they do not align properly, loosen the distributor clamping screw and rotate the body of the distributor until they are aligned. Then clamp the distributor again.

Wrench is on the screw that clamps the distributor body. Loosen it to set timing. Don't forget to re-tighten. Don't forget that the engine is running!

Remove the carburetor air cleaner and inspect it closely to see if it is clogged. The life of this component depends greatly on where you drive. If you are in doubt as to whether to replace it—do it. A clogged filter can reduce passing ability and performance. With the air cleaner removed from the top of the carb inspect all linkages for signs of binding and wear. Check the throttle valve(s) to see if they are fully open when the accelerator is fully depressed.

Check to see that the choke unloader causes the choke to open when the accelerator is fully depressed. Hold the choke closed while the pedal is being pushed down. When fully depressed, the choke should partially open. This allows a flooded engine to be cleared by cranking when the throttle is fully opened.

Make sure that choke linkages and valves are free and do not bind at any point. Varnish accumulations can cause binding and sticking. The affected parts can be removed and cleaned in carbu-

When the engine is completely cold, the automatic choke should close the choke plate as shown here. That makes it start easily and then the automatic choke will gradually open up the choke plate.

retor cleaner or with lacquer thinner. On carb types using a line to pick up heated air, make sure the passages are intact and not plugged. Specific suggestions on how to set automatic chokes are found in service manuals.

If the engine has been dying when you pull up to a stop sign you may want to reset the idle adjustment and make the engine run just a bit faster when idling. On the other hand, maybe you think it idles too fast when you pull up to a stop, and you would like to set the idle down some. This is simple to do. All you need is a screwdriver.

Most carburetors now have two idle settings. One is hooked to the choke mechanism and causes the engine to idle quite fast when it is warming up. This one you leave alone because it can get complicated.

The warm idle adjustment is the one you can alter. Locate the little screw that is acting as a throttle stop at the end of the throttle linkage or cable. If you are in doubt, fire the engine and let it idle. While it is idling you can slowly turn the screw in or out to see if it changes idle setting. Fine. You've got the right one—but don't set idle now. Put the air cleaner back on—now set the idle. Go for a little test run. If the engine dies when you pull to a stop, set the idle speed a little higher. If the car is equipped with air conditioning you should see if the engine dies when the air is switched on. This puts extra load on the engine and will make the engine die if the idle setting is right on the brink of being too slow.

Other than setting idle speed, I would recommend the normal Saturday morning mechanic keep his hands off the carburetor. Hands that don't know what they are doing cause more problems with carbs, than the units do when left alone.

This is the basic tune-up. After a couple of sessions you'll work out your own little routine and check list that will probably include checking accessory drive belts, trans fluid level and so on. The more comfortable you feel about doing your own tune-up the more you'll want to do. Soon you'll be checking everything in sight while changing oil.

POOR MAN'S DISTRIBUTOR MACHINE

There's no need for despair if you are bucks down, looking for economy and want to check your ignition advance.

If, on a cold morning, you find the choke plate partially open or all the way open like this, you are experiencing hard starting, slow warm-up, poor engine performance on short runs to the market, and eventual problems with the battery on account of all that cranking in the mornings. A shop manual for your car should show you how to fix it, or take it to a qualified carburetor specialist. If you want to know more about choke systems—how they really work as opposed to just knowing the settings—check out *Rochester Carburetors* by H. P. Books.

Have somebody work the throttle while you check everything that moves. You'll find a screw that holds the throttle mechanism partly open for idle. Turn that screw to adjust idle speed. Sometimes you will run into throttle closing dampeners like this one. Set them by the manual. Be sure they operate properly.

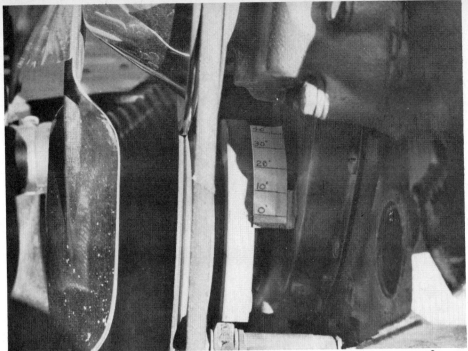

When you set ignition timing at idle speed, it will usually be between zero and 10° and the factory timing indicator shows that. If you want to check the automatic advance mechanism and plot the curve, you will run the engine faster and get more advance—up to around 20° or more in some cases. To check that, extend the range of the stock timing indicator marks by manufacturing your own scale.

First extend the measuring range of the crank damper or timing plate. This will vary from engine to engine. You pick the method that works best for your engine. Let's say there is a single pointer on the engine and the damper is marked with lines up to 10 degrees. Measure how far apart the lines are and duplicate a set of lines on a piece of tape to stick on the damper. In other words, you are degreeing the damper further—say up to 20 degrees. Disconnect the vacuum advance hose on the distributor. Connect a timing light and a tachometer. Now with a buddy reading the tach, make note of the timing as you slowly bring the revs up. Start with idle—usually about 600 RPM. Move to 1,000 RPM and then on up in 500 RPM steps to about 3,500 RPM. He calls out the RPM and you call out the advance at the crank damper. Your scribbling on a piece of paper might look something like this:

Degrees Advance	RPM
6	idle
6	1000
7	1500
12	2000
15	2500
15	3000
16	3500

CAUTION—Do not run the engine up to higher speeds without disconnecting the fan belt. The chances are extremely remote, but a fan blade can come off and become a dangerous flying object. Never rev an unloaded engine to top RPM's.

The table tells you where the advance comes in. In the example above, nothing much is happening until 2000 RPM. At that point the weights start to move out and the springs start to stretch. If you are tuning for economy, you'll want to experiment with getting the advance to come in earlier.

There are several ways of doing this.

Let's start with the springs in the advance mechanism. They can be replaced with lighter ones, or the existing ones can be stretched. To stretch the springs, remove them from the distributor and just start to insert a screwdriver between the coils, then pull it back out. Reinstall the springs and check timing again. Tweaking the spring just the smallest amount is enough to bring the advance in several hundred RPM sooner. Keep stretching with the screwdriver until you get the advance to come in between 800 and 1000 RPM. Tweak each spring. The same result can be accomplished with different weights. Naturally, a heavier weight will swing out and advance the distributor quicker. So you can get a lot heavier weight and start grinding off metal. Or you can add metal to the existing weights or you can buy a kit of springs and weights at a speed shop that will allow you to experiment with advance until your eyes are blinking like a timing light. It's worth it though on most '71 and later distributors.

TUNE-UP TOOLS

For less than a hundred dollars you have what you need to do a tune-up. Regular tune-ups and maintenance will

For this procedure you need to know engine RPM. Use your incredible all-purpose versatile Engine Analyzer instrument with the switches set for *tachometer*.

After you find out the advance curve as it is, you may decide to use the advance at a lower RPM. One way is to slightly stretch the springs on the centrifugal advance mechanism. Then recheck. Too much stretching will ruin the springs.

If your interests lead you to fancier tuning than described here, you'll need a vacuum gage. Their use is described in shop manuals and other heavy info on tuning. Also, as described in the driving technique sections of this book, a permanently-mounted vacuum gage on the instrument panel is a good idea. Either way, look around on the intake manifold and you'll find a place to connect the hose.

save you that many times over. The tools should last you a lifetime. They are available from general merchandise stores such as Sears or from automotive parts stores.

To do a tune-up correctly on a modern car you need a timing light and a dwell meter. This last item has gotten pretty sophisticated in the past five years. Dwell meters are normally called engine analyzers or some such and they are combined with ammeters, volt meter, ohmmeters and tachs. This is all to your benefit since it allows you to learn a lot more about your engine. With the instructions packaged with these analyzers you can check alternator output, regulator condition, dwell, points, accessory current draw, starter system resistance, charging system resistance, locate shorts, check alternator diodes and spark plug cables. You should have one if you plan to tune your own engine.

Here's a list of instruments and tools that are essential or handy for tune-ups. Don't rush out and buy everything on this list. Work into it gradually. Shop around enough that you know what prices are and what you get for your money. It helps a lot if you can get an experienced hand to advise you or let you watch while he uses his instruments and tools.

Timing Light—Get a "power" timing light which means that the instrument draws electrical power from a source besides the spark plug. Some hook onto the vehicle battery for power, some plug into the AC outlet on your garage wall. For most purposes the type that draws electrical power from the vehicle battery is fine. Unless you intend to work on things that don't have batteries, such as some motorcycles and similar playthings.

Dwell Meter or Engine Analyzer—As mentioned earlier, some of these include everything but a TV and tape player. Shop for what you need. The basic requirement is a dwell meter for the basic tune-up.

Tachometer—When you go a little beyond the basic tune-up you'll want a tach. May as well buy it as part of the Dwell Meter.

Vacuum Gage—If you advance beyond the basics, you'll also want a vacuum gage. Even if you have one in the driver's compartment, it's handy to have another one on a hose that you can look at while adjusting something.

Compression Gage—You will need this about two times a year. If you are on a tool-lending basis with a buddy, borrow his. You should naturally be prepared to lend him something of value from your inventory—such as a shovel.

Remote Starter Switch—Sometimes you can buy sets of tune-up equipment packaged in a flashy box as big as a pool table. Nestled here and there among all the flash will be instruments. One will be a remote starter switch. Check the package deals because sometimes the remote switch is like free.

Spark Plug Wrench—Best is a special spark plug socket that matches your socket set. These are deep enough to fit over plugs, thin enough to get into tight spaces, and usually have a built-in holder to hang onto a loose plug so you don't drop it under the car. Most, but not all spark plugs, are 14mm size and require a 13/16" plug socket. If you don't know for sure what size socket you need, outsmart the situation. Read the brand and type number on the spark plug—such as Champion J10Y. Then go to the parts place or auto store and ask for a plug socket to fit that exact plug. The man may smile but deep inside he will respect your pure logical mind.

Spark Plug Gage and Gap Setter—Buy it at the same place. You will have impressed the man so much, he will sell you a good one.

Set of Feeler Gages—Same deal.

Miscellaneous wrenches, sockets, pliers and hand tools complete your arsenal.

PARTS

All of the small parts you need for a tune-up are readily available. The parts counter at a dealership will sell an individual parts and anyone can walk into an auto parts house—they're listed in the Yellow Pages. Large department stores often stock parts needed for a tune-up. Points, plugs, filters and ignition wires, oil and transmission fluid are even sold in some grocery stores. Service stations and garages also try to keep parts on hand.

If you need an item like a fuel pump or water pump, a dealership or an auto parts store can normally take care of you—or they can tell you where to get the needed item. Most guys working behind a parts counter are very helpful and knowledgeable about the parts they handle. If you need a fuel or water pump, be ready to tell the counterman the brand name and model of the car you have, the year, the engine size—and don't be surprised if he asks you if the car has air conditioning. You see, he is looking in a great big book trying to find the right part number and he is trying to narrow it down.

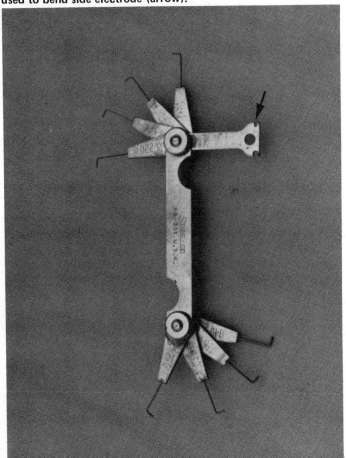

Round-wire spark plug gap gages like this are best. Notice tool used to bend side electrode (arrow).

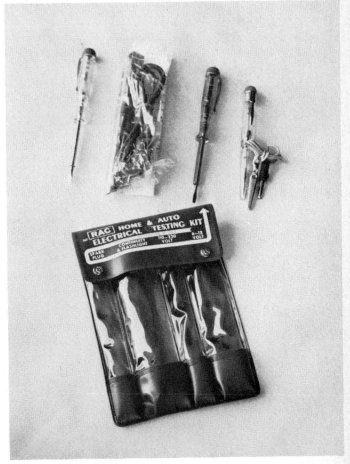

As you build your collection of tools, you'll add little things that come in handy. This kit contains electrical testers for everything except lightning.

Does timing affect gas mileage?

Ignition timing as a method of improving economy has been discussed in every garage and gas station since Henry sold his first Model T. The only hard and fast rule developed from all this discussion is each and every car/engine combination must be treated separately when it comes to tailoring spark-advance curves for economy. The manufacturers usually don't do a bad job of matching the pieces they sell you for a reasonable amount of economy. They do have to make the engine last for a period of time—at least through the warranty period. To do so the car must be capable of running on the worst gas available, driven by a lead-foot idiot while towing a 30-foot house trailer up a hill.

Conservatism, in the interest of saving the engine, is the golden rule of the manufacturer—and rightly so. It is up to the individual driver to analyze the car and his driving habits and make changes that will better adapt the car to his driving habits and make changes that will better adapt the car to his style of driving without causing durability problems.

Ignition timing consists of three parts—the mechanical or centrifugal advance curve, the initial or static advance, and the vacuum advance. Mechanical advance is determined by weights and springs inside the distributor. As the distributor spins faster and faster, the centrifugal force moves the weights outward against spring tension and the advance starts. From that point the advance continues to increase until the weights hit a metal post or other limiting device. The size of the weights and the tension of the springs determines the shape of the advance curve and its starting point, while the stop limits the total advance to something the engine can stand for prolonged high speed driving.

Initial advance is set by rotating the distributor housing when the engine is tuned up and then locking it into place. This is a constant in the overall timing picture. The engine never sees less than the initial advance and any mechanical or vacuum advance is added to the initial.

Vacuum advance is usually accomplished by a vacuum diaphram mounted on the distributor. One side of this diaphram is attached directly to a plate inside the distributor upon which the points are mounted. The other end of the diaphram is connected to a vacuum source on the engine.

Vacuum advance is the economy advance for an engine. When an engine is cruising, the manifold vacuum is high and the amount of mixture entering the cylinder is less. Therefore the pressure in the cylinder is less. The molecules of gas and air in the cylinder at high manifold vacuum are farther apart and, because they are, the flame front traveling from the spark plug can't move as fast as it would if the particles were closer together. The vacuum advance is designed to start the flame *sooner* in the combustion cycle so this less dense mixture has *more time* to burn and create a maximum amount of downward force on the piston.

However, too much advance of any kind starts the mixture burning too soon and the resultant downward force on the piston comes when the piston is still moving upward. This "force against force" is spark knock or detonation and it will ruin an engine if continued. The practice of "cranking in more advance to increase economy" should be tempered with this knowledge. Detonation can be very expensive. Just when to stop "cranking in advance" varies among engines, driving conditions, and vehicles. But if you hear pinging or knocking sounds, you've gone too far.

For this test we used our 454 CID V-8-powered Chevy truck that weighs in at a hefty 6060 pounds empty. The axle ratio and tire size combined to keep the engine turning only 2250 RPM at 60 MPH. Initial advance was varied upward from 10 degrees BTDC (Before TDC) to determine exactly the advance needed for maximum economy.

Economy increased very slightly as timing was increased from the stock 10 degrees to 14 degrees, but after that it took a nose dive. At 18 degrees economy was down as much as 9% from the maximum economy at 14 degrees. The engine kicked back against the starter and displayed traces of detonation, noticeable on some light accelerations.

Here was a case where the factory advance was hard to beat.

Just to find out how effective vacuum advance is in improving the economy of this engine, economy checks were run with and without the vacuum line connected. The results show a substantial decrease in economy. If the factories did not have this tool for tuning, we would have had our first fuel crisis in 1958.

On our shop truck, we tested different spark-advance settings to see the effect on fuel economy. The stock setting at idle is 10° which was the baseline for the test. A small improvement was obtained by increasing the advance to 14° at idle speed, but economy took a nosedive at higher settings such as 18°. Many people think more spark advance automatically gives you better gas mileage. You may find a small improvement over the factory setting if you take the time to test and find out. If you just crank in more advance, you may make it much worse.

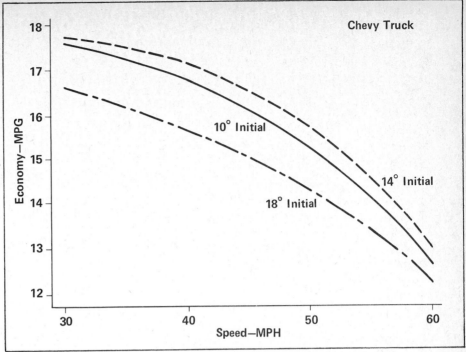

It's the vacuum-controlled spark advance that gives fuel economy in ordinary driving. To demonstrate this, the truck was tested with best spark adjustment at the distributor (14° at idle) and then the vacuum hose was disconnected for the second test. Without vacuum advance, miles-per-gallon dropped about 20% as you can see.

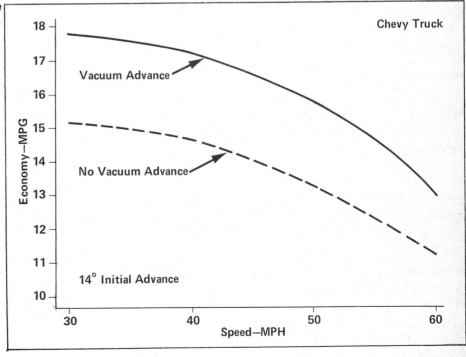

10 INSTRUMENTATION FOR ECONOMY

At the Union 76 Fuel Economy Tests, the cars were fully instrumented and the test results were scientifically checked and computed. The ordinary person can't go to this extreme, but you can do a lot with less sophisticated equipment.

There are some devices designed to show you what kind of mileage you are getting as you drive. The thinking behind this is that if you can see what poor mileage you are getting, you'll either drive less or drive in a more economical way.

Super accurate methods of measuring how much fuel is being consumed do not come easy—there are precision flow meters and the price tag reflects the quality. Laboratory and industrial flow meters can cost in the neighborhood of two grand. Obviously that's a little on the steep side for one instrument.

There are at least two miles-per-gallon meters on the market which merit a closer look because the pricing is modest—about forty bucks. The MPG meters are the same in principle—varying only in the hardware. They both come in kit form and both consist of three major parts— direct reading MPG meter, fuel flowmeter and a pulse generator that plugs into the speedometer. Actually, it's very simple and the principle has been around for a very long time—but now manufacturing procedures and hardware availability have allowed the cost to get within reason.

By definition, miles per gallon means distance traveled in a certain time divided by gasoline volume consumed in the same time. These MPG meters measure continuously as you drive and tell you at every instant how you're doing. One of the units is manufactured by Aviatric Ltd., Box 7, Romsey S05 8XS, Hampshire, England. The other is manufactured by Spacekom, Inc., of Goleta, CA and is sold by Sears. I checked on the Spacekom unit and found instructions that made sense and some easy-to-install hardware. If you really want to make a neat job of it though, don't rush. There's a lot of wiring to route—or hide—depending on how you want to handle it. Spacekom says their instrument is accurate within 3 percent from 4 to 30 miles per gallon. These MPG meters should not be confused or even compared to a vacuum gage. Vacuum does not enter the picture in either of the units.

VACUUM GAGE—POOR MAN'S ENGINE ANALYZER

A vacuum gage on your instrument panel is one of the most readily available, low-cost, useful tools if you are really serious about getting the best possible

Instruments such as this SpaceKom unit measure miles-per-gallon and display it all the time you are driving. You can see the effect of your driving technique and the effect of good maintenance and tuning. They are relatively inexpensive and not difficult to install.

Some cars now come equipped with a built-in vacuum gage. This one is called a fuel-economy gage but it reads intake manifold vacuum just like uneducated vacuum gages do. When you mash on the throttle, it scolds you by pointing to *MINIMUM FUEL ECONOMY.*

At your local friendly auto store, you can buy an add-on vacuum gage and install it in your own car. Not as fancy as the MPG meters or the factory-installed economy-meters, but it will put bucks in your wallet if you pay attention to what it says.

mileage from your car. Buy one, install it and live with it. Look at it often.

Simply stated, high vacuum means better mileage than low vacuum. Try to drive with the vacuum gage holding a steady reading or varying smoothly. You are attacking the task the wrong way if you roar away from the stop light with the vacuum gage needle slammed down to zero, then let off suddenly and smile as the gage yanks over to 20 or more. It's far better to ease away with a reading of 10—lotsa luck—then drive steadily with the gage sitting at 15.

A vacuum gage measures vacuum in the intake manifold or in a particular part of the carb circuitry—depending on where you connect it. Most vacuum meters are marked in inches of mercury (Hg) from 0 to 30. Some have various colored bands labeled *poor, good, excellent, late timing, bad plugs,* etc. This is useless information for the most part. The needle position indicating something amiss or excellent mileage will vary from one engine to another. The main thing to remember is: Keep the needle steady as you drive. The gage meters your driving habits—pay attention!

A vacuum gage is extremely easy to install—normally there are no holes to drill—it's that simple. A vacuum gage can be hung under the dash or taped on the steering column—according to how particular you are. Depending on the brand of gage and the car it is used on, some gages will have a "needle flutter" or bounce at idle that begins to smooth out as the vacuum increases. This flutter can be eliminated by reducing the size of the orifice through which the gage gets vacuum. There are a lot of exotic ways of doing this—but the simple way is to put a small plug of modeling clay up into the fitting on the back of the gage. Now punch a hole in the clay with a large needle. The hole needs to be around .020 inch in diameter to damp the needle sufficiently.

Vacuum gages are available at most automotive parts stores, discount houses, and through auto mail-order firms. They are inexpensive—less than $10.

There is normally a fitting on the intake manifold where you can connect the hose.

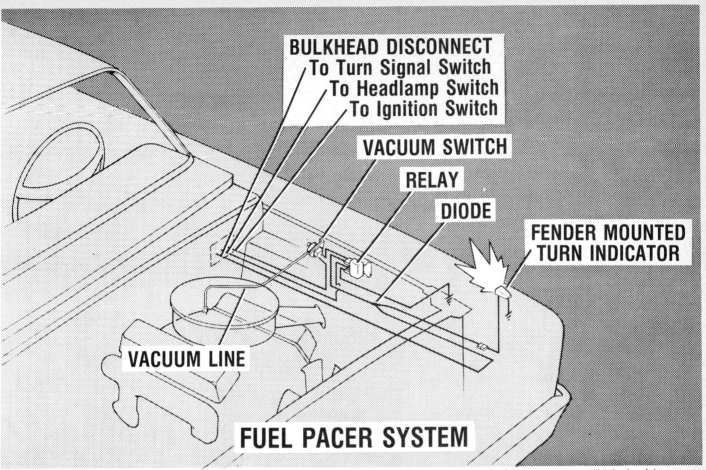

Chrysler has developed instrumentation which they call the FUEL PACER system. Instead of a vacuum gage which the driver must remember to watch, it uses a vacuum-operated switch to turn on a light in the driver's field of vision.

INSTRUMENTATION

Chysler Corporation has a neat little device called a Fuel Pacer which alerts a driver when he's using more fuel than necessary.

It actuates a light on the left front fender whenever the carburetor begins to enrich the fuel mixture for an increased power demand. If the light goes on, the driver sees instantly he should drive the car at a more efficient operating level.

Rapid starts and accelerations dip into the power reserve. With the aid of the Fuel Pacer, drivers can develop the right touch, using much the same principle that fuel-economy test drivers use, yet emergency power is always available when needed. No price has been announced although it is expected to be in the $10 to $12 range.

The part number for the system is 3740920 and although the system is designed to be used on Chrysler Corporation cars, it can be modified to work on most other cars.

The Chrysler system of activating a light may prove to be far more beneficial in encouraging prudent driving habits than vacuum gages and other such devices which must be constantly monitored. These gage-type devices might be called "passive" as opposed to the "active" system of a direct warning when excessive fuel is being used.

This chart shows that the Pacer Light is turned on whenever the driver uses enough throttle to turn on the power system in the carburetor. These drawings are courtesy of Chrysler Corporation.

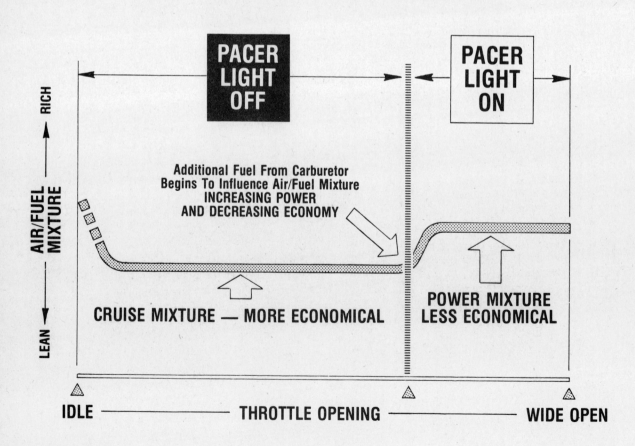

This is the ultra-accurate Autotronics Fuel Flow Meter set up with a cool-can to minimize the effect of temperature changes on the volume of fuel. We kept the cool-can filled with ice during all our testing.

THE FUEL COUNTER

There are a lot of methods I could have used to measure the amount of fuel consumed in all the tests we ran. But none is as accurate and convenient as the Autotronics Fuel Flow Meter, available from Autotronic Controls Corporation, 6908 Commerce, El Paso, Texas, 79915, (915) 772-7431. It is the most accurate fuel flow meter available for in-car use at the time this is written. Just how accurate is it? It gives a readout of fuel consumed every time the engine drinks 1/1000 of a gallon, and that ain't much gasoline!

Lines running from the fuel tank bring gas into the meter, which keeps track of the gas as it flows to the engine. The unit contains an electric fuel pump so pressure against the measuring part of the meter is constant. There's a reset button on the instrument so the readout gage can be reset to zero for each run and there's a remote button held by the driver to start the count. A plenty slick device. Unfortunately, at a cost of $2400 each, everyone interested in fuel economy can't have one.

Autotronics fuel meter, cool can and other instrumentation in one of the test cars used to gather data for this book.

THE WORLD'S MOST ADVANCED COMPUTER

There's an old joke. "What's the most advanced, adaptable, programmable, computer-controlled, non-linear servo-mechanism that can be produced by unskilled labor?" The answer is, "People."

We have to get serious about conserving fuels and energy because the world is running out of the fossil fuels we have been using.

That means we have to use the computer in our skulls and re-program it from some of the old ways. We have been using automobiles as extensions of the ego and a lot of other things besides a way to keep from walking. Nine drivers out of ten have an overwhelming urge to pass the car in front—just because it's in front. We have to stop that.

All of the economy tuning, tricks, flashing lights, gages and federal laws won't help as much as some very simple things. Drive less when you can and drive conservatively when you do. Drive the minimum car that suits your needs. Keep it tuned and adjusted so it doesn't waste money and fuel today that you may wish you had tomorrow.

This book has a lot of information for the average driver—some of it common knowledge and common sense, some of it less well known but nevertheless useful. I hope the information serves you well and gets you started toward the goal of economy—for yourself and for all the rest of us who worry about money in the bank and gasoline in the tank.

Doug Roe Engineering served as consultant on development of this Bricklin sportscar prototype. The same basic ideas and test procedures are used on a custom high-performance car as on your family car. The Autotronics meter is part of the instrumentation.

Lines running along the left side carry fuel to the flow meter and back to the carburetor. Hose on right side is vacuum. The Pinto survived the operation and lived a long and happy life thereafter.

A professional automotive tester must have the mental concentration and abilities of a one-man band. This driver operates the car, watches the speedometer, holds the taped-on vacuum gage steady, works the stop watch, and uses his right thumb to start the fuel-flow meter. Besides that, he listens to the radio and chews gum!

NOTICE: The information in this book is as nearly correct and complete as we could make it—based on our own tests and experience. All recommendations and instructions are presented without guarantee, although in our opinion they are valid. Because products, vehicles and parts change and because methods of application are beyond our control, publisher and author expressly disclaim liability for use of any or all data and instructions in this book.

The only simple and inexpensive way to make precise fuel-quantity measurements is to do it "by hand" using a graduated burette available from chemical suppliers and school-supply businesses.

Writing a book and illustrating it is no light task. I want to thank those who helped get it together. Jim LeRoy, a long time friend and currently a business partner, did much of the road testing and preparing of graphs which illustrate the results. Jim is in photo above. John Thawley, famed auto writer, did considerable research for several sections and snapped a good number of the photos. Carl Shipman, Nancy Fisher, Lou Duerr and Josh Young are part of the H. P. Books Staff, who put this material into book form. If you find the book helpful and informative, it's because a lot of good talent had a hand in it.

Doug Roe